Reasoning Across Domains

European University Studies
Europäische Hochschulschriften
Publications Universitaires Européennes

Series XX
Philosophy

Reihe XX Série XX
Philosophie
Philosophie

Vol./Band 693

PETER LANG
Bern · Berlin · Bruxelles · Frankfurt am Main · New York · Oxford · Wien

Harry Witzthum

Reasoning Across Domains

An Essay in Evolutionary Psychology

PETER LANG
Bern · Berlin · Bruxelles · Frankfurt am Main · New York · Oxford · Wien

Bibliographic information published by Die Deutsche Bibliothek
Die Deutsche Bibliothek lists this publication in the Deutsche Nationalbibliografie; detailed bibliographic data is available on the Internet at ‹http://dnb.ddb.de›.

British Library and Library of Congress Cataloguing-in-Publication Data:
A catalogue record for this book is available from *The British Library*, Great Britain, and from *The Library of Congress*, USA

I would like to thank all those who helped me in completing this work: Dr. Stephen Laurence for his valued comments and criticisms on innumerable drafts and Dr. Robert Hopkins for his kind help and comments on the book. I would also like to especially thank Prof. Dr. Peter Carruthers for always inspiring and insightful discussions and comments in the first stages of the book and Dr. George Botterill for his comments on the initial chapters. I send out my gratitude to my fellow students for providing me with many ideas both in seminars and private conversations – many are the influences I received during the writing of my book from supervisors, professors, fellow students, and friends. In addition, I have greatly profited from the regular workshops of the Hang Seng Centre for Cognitive Studies based at the University of Sheffield, which provided us with an exceptional interdisciplinary context for exchanging ideas and research.

My deepest gratitude goes to C.S. for her constant support, motivation, and trust in me.

ISSN 0721-3417
ISBN 3-03910-978-2
US-ISBN 0-8204-8365-6

Table of Content

Introduction

> In the distant future I see open fields for far more important researches.
>
> Psychology will be based on a new foundation, that of the necessary acquirement of each mental power and capacity by gradation. Light will be thrown on the origin of man and his history.
>
> Charles Darwin (1859/1998, pp. 367-368)

When psychologists refer to the "mind", they mean the totality of cognitive capacities, realised in the brain, that are responsible for all conscious and unconscious mental events, and that control all behaviour. Cognitive capacities, however, can differ vastly from one another, and are as varied as our ability to choose an attractive mate out of a crowd, our ability to selectively attend to spoken words in the stream of background noises, and our ability to see three-dimensional objects in colour. The objective of psychology is to identify each cognitive capacity, to study its nature, and to explain how it works individually and in relation to other capacities.

Recently a debate has erupted in psychology about the best ways to approach the explanation of cognitive capacities. At the core of this debate are questions about the advantages or disadvantages of applying an evolutionary perspective in psychology. Traditional cognitive psychologists – those who dismiss the evolutionary perspective – pit themselves against evolutionary psychologists – who whole-heartedly champion it. And the stakes for psychology are high, since application of the evolutionary perspective has ramifications for the cognitive structure of the mind.

A common view among traditional cognitive psychologists is that the mind resembles a blank slate virtually free of specialised cognitive content. Humans are treated as though they were biologically empty, except for small numbers of general-purpose instincts. Human cognitive equipment is thought to be almost entirely the product of culture that inscribes cognitive content onto the mental slate via learning or teaching. An old tradition within philosophy has it that there is nothing in the intellect that was not previously in the senses except the intellect itself. This intellect is understood as a general-purpose cognitive mechanism that is content-independent in a strong sense. According to this view, humans apply the same general-purpose cognitive mechanisms in a wide range of different domains. Thus the same general-purpose cognitive mechanisms are thought to govern, for example, how one learns a language, how one learns to choose the right mate, and how one learns to support one's kin. This is because the general-purpose cognitive mechanisms are assumed to operate uniformly, regardless of the contents they are operating on. That is also the reason why cognitive content has to be derived from the external physical or social environment.

Moreover, the cognitive specialisation observed in human adults is claimed to be the product of a developmental process in which content-independent general-purpose cognitive mechanisms build up specialised cognitive content from experiences given to them during learning. Traditional cognitive psychologists, while not denying that the mind is an evolutionary product, typically claim that the evolution of a culture-absorbing mind like ours was accompanied by the erasing of whatever specialised circuitry the brain might have had before. Such considerations therefore lead traditional cognitive psychologists to conclude that the evolutionary perspective in psychology will not benefit greatly our understanding of the mind.

Evolutionary psychologists are diametrically opposed to the above understanding of traditional cognitive psychology. Evolutionary psychologists claim instead that the cognitive capacities, which comprise the mind were formed and shaped by natural selection to solve specific problems concerning survival or reproduction that our hunter-gatherer ancestors faced. Moreover, they claim that there is a degree to which natural selection among our ancestors still influences our modern human behaviour: that is, our modern skulls still house a "stone-age" mind. According to evolutionary psychology, the human mind, far from being the blank slate which traditional cognitive scientists view it, is brimming with specialised cognitive capacities that initially helped humans to solve the adaptive problems from an early age. For each domain that gives rise to an adaptive problem there corresponds a content-dependent cognitive capacity that was designed by natural selection to fulfil its problem-specific task. It is therefore not true that the same cognitive mechanisms are applied uniformly in a wide range of different domains. For example, different cognitive capacities are thought to govern how one learns a language, how one learns to choose the right mate, and how one learns to support one's kin.

In addition, evolutionary psychologists are typically suspicious of general-purpose cognitive mechanisms in principle. They claim that a correct understanding of the evolutionary process shows that general-purpose capacities cannot be a default position; on the contrary such general-purpose capacities will be the exception to the rule once evolution has been brought into the picture. Thus, claim evolutionary psychologists, the evolutionary perspective will greatly change how we view the mind.

Evolutionary psychologists are, therefore, opposed to traditional cognitive psychology on at least two counts. First, evolutionary psychologists maintain that the mind is largely or even entirely composed of specialised and content-dependent cognitive capacities that were designed by natural selection to solve specific problems of the physical and social world. Second, they are suspicious about general-purpose cognitive capacities and demand that such capacities must be validated from an evolutionary perspective.

The main purpose of my thesis is to critically assess the theory and arguments of evolutionary psychology. I defend evolutionary psychology against its critics and I demonstrate that a weak version of evolutionary psychology provides a promising research programme for future studies of the mind. Very broadly speaking, strong versions of evolutionary psychology differ from weak versions in virtue of the fact that strong versions do not allow general-purpose cognitive mechanisms into the mind's structure, whereas weak versions do accord general-purpose cognitive mechanisms a place within the mind's structure – provided that these cognitive mechanisms work in conjunction with the specialised cognitive equipment and that an evolutionary account of them can be given. As I said, I defend a weak version of evolutionary psychology, and I explore the general implications of this version for general-purpose cognitive mechanisms.

The thesis is structured as follows: Chapter I introduces evolutionary psychology by discussing its basic concepts and theoretical arguments. Important distinctions between elements of evolutionary psychology are made, and different versions of evolutionary psychology are discussed. This and the next chapters form an investigation within philosophy of psychology that analyses the conceptual framework of evolutionary psychology and tests it for its consistency and fruitfulness.

Chapter II presents objections to evolutionary psychology and critically evaluates these objections. The result of this evaluation is that weak versions of evolutionary psychology can successfully rebut these objections. The outcome of the evaluation is therefore an evolutionary position that views the mind as largely composed of specialised cognitive equipment, but which also includes general-purpose cognitive mechanisms as part of the cognitive furniture.

Chapter III gives a plausible account of the evolution of general-purpose cognitive mechanisms within weak versions of evolutionary psychology. The chapter clarifies why general-purpose cognitive mechanisms are made necessary by problems the mind confronts as a result of its specialised cognitive equipment. Faced by the simultaneous activation of specialised cognitive capacities the mind needs to find solutions that combine or integrate these cognitive capacities in order to deduce adaptive responses to these problems. The chapter then critically assesses a recent proposal from evolutionary psychology that aims to solve these problems by means of special-purpose cognitive capacities, and finds this proposal wanting. Finally, the chapter outlines conditions that would have been conducive to the development of general-purpose cognitive mechanisms in non-human species. This last outcome validates general-purpose cognitive mechanisms from an evolutionary perspective, and weak versions of evolutionary psychology are thereby identified as promising research strategies for investigations of the mind.

Chapter IV then examines evolutionary psychology's core assumption about the evolutionary study of the human mind. If the human mind is a product of evolution, then one should expect its cognitive structures to have been developed by small and incremental changes from pre-existing cognitive structures in non-human animals. Evolutionary psychology's core assumption therefore is that non-human animals possess cognitive structures. I claim that for a complete evolutionary understanding of the human mind, what is therefore necessary is a systematic ordering of the cognitive options that evolution provided for non-human animals. This chapter develops and defends an evolutionary taxonomy of cognitive capacities in non-human animals on the basis of an analysis of an influential model proposed in the literature.

Chapter V assesses arguments from philosophy that challenge evolutionary taxonomies such as the one defended in the previous chapter. According to these arguments, non-human animals cannot properly be said to possess beliefs and other mental states. I argue that these arguments from philosophy are defective, and that we are justified in claiming that non-human animals possess mental states of some sort. The outcome of the chapter is that evolutionary projects aimed at understanding the mind do not succumb to these last sorts of arguments, and thus that we can confidently continue evolutionary projects.

The main conclusion of this thesis is that weak versions of evolutionary psychology, as I develop them here, do provide a promising research strategy in psychology and should thus be taken seriously by anyone interested in the mind and its cognitive structure. Weak versions of evolutionary psychology are not only able to integrate findings showing that the mind contains large amounts of specialised cognitive capacities, but they are also capable of outlining evolutionary accounts of the origin of general-purpose cognitive capacities. Together with the accounts of the evolutionary taxonomy of cognitive capacities in chapter IV, and its defence against arguments from philosophy in chapter V, weak versions of evolutionary psychology stand as a serious alternative to traditional cognitive psychology.

I. Evolutionary Psychology: Concepts and Arguments

Evolutionary psychology is a new research approach to psychology. It is not a domain of study like linguistics, vision science, or social psychology. Rather, it is a way of thinking about psychology that is applied to any topics studied within psychology. An important role of evolutionary psychology is to reflect the basic concepts of cognitive psychology and to introduce suggestions on how to improve these concepts. A very important concept of cognitive psychology is the concept of a *cognitive architecture*. I want to describe this concept before introducing the details of evolutionary psychology. Let me start with a little sci-fi.

The sci-fi series *Star Trek: The Next Generation* introduced us to lieutenant commander Data, an android life form. Data visited the elite Starfleet Academy, took his degree there, before joining the crew at Starship Enterprise where he took over a command position. His creation was made possible by recently developed technology: the positronic brain matrix (whatever that means). Data is a complex and sophisticated automaton, possessor of a full-blown natural language and flexible learning mechanisms. While surpassing humans in physical strength as well as thinking or computing power, he does not possess the ability to experience either pain or pleasure – indeed any emotions whatsoever. However, he still has the undying urge to become human.

The creation of such automata would spark debate at conferences around the world, but the truth looks much bleaker. The fact is, we have not even managed to build simple robots to help us wash the dishes and bring us tea when prompted to. For sure, there are simple artificial devices that, e.g., imitate the behaviour of ants, but nothing we have can compare with the behavioural abilities of even rats or other lower mammals. It is therefore trite to remark that such androids as Data belong to the realms of science fiction, as of yet. But why is that so?

Among the many obstacles that hinder our attempts to create such artificial life forms is our lack of a comprehensive understanding of the basic structure and organisation of the human mind, which we could implement in real hard- and soft-ware. In other words, we lack a good understanding of the cognitive architecture of the mind. A cognitive architecture of the mind is a complete survey of all the cognitive systems housed within our minds, their functional capacities, and their relation to each other.

We may initially think that the acquisition of such a survey of the mind's cognitive systems can proceed easily enough by introspection (Titchener, 1912; Wundt, 1897). But such an assumption would be rash.[1] For cognitive architectures are not properties of organisms that can easily be investigated by probing our intuitions. Cognitive architectures do not lie at the surface, so to speak, ready to be reported by psychologists with a trained introspection. Many of the processing feats that our minds perform in their encounters with the world are simply not accessible to conscious introspection (Buss, 1999; Pinker, 1996). Many computational processes are performed on a sub-personal level (Stich, 1978), and it is only the output of these processes that may enter into consciousness and be accessible to introspection. So, for instance, we do not have direct introspective access to all the operations that produce our image of a three-dimensional world of objects on the basis of a two-dimensional stimulus configuration that strikes our retina. This is why we need a science of vision (Marr, 1982). Generally speaking, introspection will not be the most reliable guide if we want to acquire a detailed survey of all the cognitive processes and an understanding of how they work. For introspection may oversimplify the underlying processes and it may also distort our project of understanding our cognitive architecture (Cosmides, and Tooby, 1994a).

1 For an early refutation of the methodology of introspection in psychology see Dunlap (1912) and Watson (1913).

With introspection disqualified as a reliable methodological tool for investigating the cognitive architecture, cognitive psychologists turn to other powerful methodological tools that may guide them in their quest for the natural joints of the mind. Contrary to introspection, such methodologies must thus be theoretically motivated and cannot be given to us by intuition alone. All cognitive psychologists actually employ such theoretically informed methodologies to guide their experimental designs, but it is evolutionary psychologists' main empirical tenet that evolutionary theory will lead psychologists to a better and more rigorous understanding of the cognitive design and organisation of the human mind.

To understand better the impact of evolutionary psychology on the understanding of the mind's structure, I shall now introduce evolutionary psychology in more detail. Section 1. introduces the core concepts that organise the field of evolutionary psychology and section 2. presents the theoretical arguments that evolutionary psychologists use to support their particular views on the mind. In this chapter I will not endorse the arguments, but merely present them as evolutionary psychologists use them to support their own claims. The evaluation of these theoretical arguments will be the topic of the next chapter.

1. Core concepts of evolutionary psychology

In this section I shall present the core concepts that define the field of evolutionary psychology. Evolutionary psychology is a new approach in psychology that has gradually taken shape over the last two decades (Barkow, Cosmides, and Tooby, 1992; Buss, 1999; Pinker, 1996). Its goal is to integrate cognitive psychology with recent theoretical advances in evolutionary biology, cognitive science, and anthropology to increase our understanding of the structure and organisation of the mind. As *psychologists*, on the one hand, evolutionary psychologists are interested in developing and testing hypotheses about what kinds of cognitive machinery our mind contains. The aim is to arrive at a complete list of all the cognitive systems that one can find within the mind. As Leda Cosmides and John Tooby write,

> Just as one can now flip open Gray's Anatomy to any page and find an intricately detailed depiction of some part of our evolved species-typical morphology, we anticipate that in 50 or 100 years one will be able to pick up an equivalent reference work for psychology and find in it detailed information-processing descriptions of the multitude of evolved species-typical adaptations of the human mind, including how they are mapped onto the corresponding neuroanatomy and how they are constructed by developmental programs. (Cosmides and Tooby, 1992, p. 68)

As *evolutionists*, on the other hand, evolutionary psychologists have a particular way of deriving those hypotheses about the cognitive machinery. Their distinguishing approach is to develop hypotheses about the human mind by assuming that it is a product of evolution and thus subject to the logic of the evolutionary process. The mind, according to evolutionary psychology, is an ancient engineering product that evolution designed specifically to survive and reproduce in ancestral environments.

This general consideration is the common credo of evolutionary psychologists. But as soon as one wishes to flesh out the position of evolutionary psychology in more detail, one quickly runs into problems. For evolutionary psychology is not free of substantial but healthy debate. Even among evolutionary psychologists there is considerable debate about basic research strategies. Consequently, a common position that all evolutionary psychologists may wish to subscribe to is not always easily discernable. Nonetheless, the following may be taken as something like a lowest common denominator. All evolutionary psychologists subscribe to at least the following theses:

1. Cognition is a kind of computation.
2. Minds are largely modular in nature.

3. Modules are biological adaptations produced by natural selection.
4. Evolutionary theory is a good heuristic device for designing experiments to track down cognitive systems.

An explanation of these theses as evolutionary psychologists use them is now in order.

1.1 Cognition and Computation

Evolutionary psychologists subscribe to *the computational theory of mind.*[2] The computational theory of mind is currently the most influential model of the mind in cognitive science. It views the mind as implementing a computer that receives input from its various sense organs, stores it as representations in specialised memory systems, and manipulates representations in accordance with computations. The characteristics of computations involve formally specified rules that access and manipulate representations on the basis of their formal properties alone.[3] Computations thus necessarily determine the nature of mental representations.

Representations, according to the computational theory of mind, are generally believed to possess semantic as well as syntactic properties in very much the same way that public language expressions do. The syntax of a mental representation is given by a formal specification of the representation that disregards all semantic

2 For a detailed discussion of the computational theory of mind see especially Field (1978), Fodor (1975; 1983, 1987), Sterelny (1990), and Stich (1983).

3 These rules might either be explicitly represented within the system in such a form that makes them accessible or manipulable by the system or they might merely be implicitly represented within the basic operating principles of the system. In the latter case the system does not have the ability to access them or manipulate them by computational procedures, they are rather built or wired into the design of the system. For a detailed discussion of the explicit/implicit discussion see Fodor (1975).

features it might have such as *truth*, *reference* and *meaning*.[4] In the course of computing its input, a computer is thus sensitive solely to syntactic properties of the representation. Moreover, while computations are sensitive solely to syntactic properties of representations, they can nonetheless be truth preserving as long as the syntactic properties of representations somehow mirror the semantic ones. Given that such a mirroring between the syntactic and semantic properties holds, the computer will be capable of moving from a *true* representation to another *true* representation by transforming the "shape" of the one representation into the "shape" of the other.

Applying the computational theory of mind to our everyday understanding of propositional attitudes such as belief and desire enables us to characterise these propositional attitudes in computational terms too (Fodor, 1987):

(1) For any organism O, and any propositional attitude A with a content that p, there is a computational relation R and a mental representation MP such that

MP means that p, and
O has A if and only if O bears R to MP.

So, for instance, for you to believe that gold is yellow is for you to be computationally related *in a suitable way* (that is characteristic for believing) to a mental representation that means that gold is yellow (in an appropriate medium in which the representation can be used

4 It has proven very difficult to pinpoint precisely the nature of the formal-syntactic properties of mental symbols (Crane, 1990). But for the purposes of this section we can define the formal properties of mental symbols negatively as the class of all the properties that are *not semantically* specified through such notions as truth, reference, and meaning. Commonly the mental representations are understood as having a physical "shape" analogous to the visible shapes of written sentences or the phonetic properties of uttered sentences of a natural language. The computational theory of mind holds that mental representations are somehow stored in the electrochemical states of the brain (Fodor, 1975; Dennett, 1978).

by the mind[5]). It is common in the literature to describe this *suitable way* in which an organism is computationally related to a mental representation by a metaphor of the following sort: to believe that p, for instance, is to have a mental representation that means that p inscribed in one's *belief box*.[6] A belief box is simply a short hand for a full specification of the causal-functional role beliefs generally play in the cognitive economy of the organism – i.e. typical relations beliefs entertain with perceptual information, desire states, and behaviour.

Furthermore, thought processes leading from one propositional attitude to another during mental episodes such as reasoning are given the following analysis according to the computational theory of mind:

(2) Mental processes are causal sequences of tokenings of mental representations.

So, for instance, to think that *I am thirsty, so I go to the kitchen* is for me to realise a mental representation that means *I am thirsty* that causes another mental representation that means *I go to the kitchen* in a computationally suitable way.

5 There is much controversy about the appropriate medium of the mind in which such representations are manipulated. Some researchers, e.g. Fodor (1975), think that it must be an innate medium such as a language of thought which must be presupposed, for instance, to explain how natural languages are learned. Other researchers, e.g. Harman (1973), think that the appropriate medium is a public natural language. In the following I will presuppose the former option. One reason for my decision not to follow the public language option is the reasoning that some cognitive system must already be in place to act as a medium for non-human animals and generally pre-verbal thinking (see also Fodor (1975) for similar reasoning).

6 This is a metaphor that goes back to Stephen Schiffer (1987).

As indicated above evolutionary psychology shares this particular picture of the mind with almost all of the rest of cognitive science.[7]

1.2 Cognition and Modules

Evolutionary psychologists' thesis that the mind is composed of modules is a claim about the cognitive structure and organisation of the human mind. Modules are functional subparts of the mind that work in relative isolation from each other. Moreover, each module generally uses modes of problem solving that are specific to one *domain* and that do not easily extend to other domains. A *domain* in this context is usually characterised as a distinct class of phenomena in the environment that share certain properties and that modules are dedicated to solving.[8] Such domains are, for example, face recognition, mathematical understanding, or basic understanding of physical objects (Hirschfeld and Gelman, 1994). Because the performance of a module is thus dedicated to a specific domain the module cannot be used to solve problems that differ in important ways across a wide range of domains. Take, for instance, language (Chomsky, 1988) and vision (Marr, 1982). Both faculties of the mind use distinctive specialised modes of problem solving that cannot be used to solve the other module's problems. The procedures responsible for extracting information about three-dimensional objects from two-dimensional stimuli impinging on the retina are very

7 This is not to say that evolutionary psychologists cannot rely on alternative theories of mind different from the computational theory of mind. There is no in principle ground of why evolutionary psychologists could not incorporate, for example, connectionist models of the mind in their views. The appropriateness of connectionist models depends on the types of connectionist models that are used for the task. I will not consider this option in what follows as this option is rarely mentioned or discussed in the evolutionary psychology literature.

8 The concept of a domain is sometimes defined differently. For different definitions of domains see Hirschfeld and Gelman (1994: 21).

different from the procedures responsible for the extraction of semantic information from phonetic utterances. The former belong to vision, the latter to language, and neither of these two faculties can use the other's resources to solve its own problems.

There is now strong and growing empirical support for the modularity thesis from diverse disciplines (Carey, 1994; Gallistel, 1990; Hirschfeld and Gelman, 1994). Some support comes from developmental studies. Those studies have shown that developing cognitive abilities of human infants display a high variability in cognitive sophistication within a single individual at a given moment in time (Gelman and Baillargeon, 1983). While some abilities show high-level performance very early on in development, others do not appear until much later. Psychologists have been able to trace very sophisticated cognitive capacities in infants as early as several months old. So, for instance, 6 month old infants already show precocious abilities in their understanding of basic principles of the behaviour of physical objects (Spelke, 1988, 1990, 1994): they know that objects travel on a continuous path even if they are occluded from view, that two objects cannot occupy one and the same place at the same time, and that two surfaces moving together belong to the same physical object. Other data show infants (even neonates) to be selectively responsive to features of the human face (Slater and Kirby, 1998; Slater and Quinn, 2001).

These and similar findings strongly suggest that the mind contains specialised modules that guide the developmental trajectories specific to all cognitive abilities. Modules could account for the high-level abilities infants manifest in particular domains since one characteristic function of modules is that they enable infants to access specialised modes of processing very early in development. Since modules work in relative isolation from each other, if one module goes on-line, it will give the infant the benefits of its cognitive resources. Other modules might still lie dormant in the infant's mind, waiting for the right circumstances to trigger them, thereby giving rise to the observed variability in cognitive sophistication.

This variability of cognitive sophistication in development is not something one would expect if the child used solely general-purpose learning abilities in its early development. The point of such general abilities is precisely that they can be used across a wide range of different domains [a view held, for instance, by Jean Piaget (1971)]. Armed solely with general learning abilities, the child should manifest a completely different pattern of development. Rather than showing the observed variability in sophistication across different cognitive abilities, the child would be expected to show a more linear development common to all of cognition along several phases of development (depending on the growing sophistication and power of the general learning abilities). The fact that such linear development is not observed, gives support to the claim that the mind is structured along modular lines.

Further support for the modularity thesis comes from data concerning normal adults whose brains have been damaged by strokes or accidents (Ramachandran and Blakeslee, 1998). The behaviour of brain-damaged persons sometimes manifests cognitive dissociations that disrupt particular cognitive faculties while leaving others intact. For instance, brain damaged patients may have impaired face recognition abilities while showing normal performance on other spatio-visual and general intelligence tasks. Again, the selective impairment of narrowly circumscribed cognitive faculties seemingly supports the view that the mind is structured along modular lines.

Finally, support for the modularity thesis also comes from data about abnormal patients, the so-called "savants" (Smith and Tsimpli, 1995). These patients manifest exceptionally good performance on some cognitive abilities, but are seriously handicapped in other cognitive domains. Persons affected by Williams Syndrome, a rare genetic disorder (Bellugi, Wang, and Jernigan, 1994), show exceptionally strong expressive language skills, good interpersonal relationships, and music skills. But they suffer from several intellectual handicaps: they show poor performance on fine motor control and spatio-visual tasks, and their IQ ranges between 50–70 which makes them moderately retarded. Such dissociations yet again seem to support a mental organisation that allows for some faculties to be selectively disrupted in development while leaving others intact.

While there is some consensus in developmental psychology on the modular structure of the mind, however, there is still considerable debate about the precise nature of modules and their implication for our understanding of the developing mind (Elman, Bates, Johnson, Karmiloff-Smith, Parisi and Plunkett, 1996). The bewildering number of different types of modules further complicates the debates surrounding modularity considerably. Alongside cognitive modules,[9] there are also different types of non-cognitive modules, which I will not discuss further since they are not directly relevant to cognitive psychology.[10] Some scientists, then, have distinguished perceptual modules (A) from conceptual modules (B). Others have introduced a distinction between static modules that explain given capacities at a given time (C) and dynamic modules that explain the development of certain capacities (D). All of these types of modules

9 Some scientists think that some cognitive modules are already fully specified in the infant's mind and require only the right kind of experience to go online. Development of such modules is more an affair of triggering an already pre-existing mental structure. Others, for example, Karmiloff-Smith (1992) and Gopnik and Meltzoff (1997) opt for a gradual specialisation of neural circuitry that develops during ontogeny rather than for an innate pre-specification of cognitive modules. Specialised circuitry develops largely as a function of experience and does not have to be innate from the beginning.

10 Non-cognitive modules are modules that are not essentially described in terms of knowledge-states and computations manipulating them. So, for example, neuro-physiologists posit *neuro-physiological modules*. These are functional subparts of the brain whose workings can be explained in terms of brute neurophysiological machinery. The introduction of knowledge-states and computations does not do any real work here. Another example: developmental biologists posit *developmental modules*. These are parts of the developing embryo that can develop largely independent of the context in which they occur (Wagner, Mezey, and Calabretta, in press). A final example: evolutionary geneticists posit *genetic or evolutionary modules*. These types of modules are semi-independent genetic complexes with tightly integrated gene-actions and products (Wagner, 1995).

are important for a detailed exposition of evolutionary psychology and I shall thus discuss each of them in what follows.

(A) Perceptual Modules

Every discussion of modules must take account of the classic analysis of modularity given by Jerry Fodor (1983, 2000).[11] Fodor views the mind as essentially a two-tiered system containing central cognition and input/output modules. According to Fodor, central cognition comprises the conceptual abilities that are implicated in fixing beliefs and decisions, and they are non-modular in nature. Fodor thus maintains that it is only the input and output systems realised in sense organs and motor systems that are modular in the strictest sense. Those modules tend to possess the following correlated properties (Fodor, 1983):

1. Modules are domain-specific in as far as they use specialised bodies of knowledge or computational mechanisms that are devoted to solving problems that arise within specific domains. Moreover, particular modules cannot properly contribute to the solution of problems arising outside their specific domains.

2. The operations of modules are mandatory. Once an appropriate input activates a module, the module's operation cannot be inhibited. Processing is executed automatically.

3. Modules work in relative isolation from each other. The resources contained within one module are closed-off from other parts of the mind. There are two ways in which modules can be said to be isolated from each other (Botterill and Carruthers, 1999):

11 But see also Chomsky (1988) and Garfield (1989) for early discussions of modularity.

(i) A module can be isolated from other parts of the mind because it grants them only *limited access* to the specialised resources hidden within it. It follows that even if other parts of the mind could benefit from the specialised resources of the module, they cannot use the resources because they are not given access to them.

(ii) A module can be isolated from other parts of the mind because it is *informationally encapsulated* from the rest of the mind in so far as information stored in other parts of the mind cannot influence the computations within the module. It follows that even if other parts of the mind contain information that could benefit the module, it cannot use the information because the module is not given access to them.[12]

4. The computations of modules are fast, since there is no interference by other systems external to a particular module due to its relative isolation from other parts of the mind.

5. Modules deliver shallow output, i.e. non-conceptual information, which needs to undergo further processing somewhere down the line.

6. Modules are associated with fixed neural architecture in the brain.

7. Modules exhibit specific break down patterns. They can be selectively impaired in development or adult stages while leaving other systems unimpaired.

12 Informational encapsulation is not the result of performance limitation. Even if parameters such as motivation and attention are controlled for the feature of encapsulation would still prevail because it is an architectural feature: a feature that is instantiated by the structure of the mind itself. Put differently, we might claim with Scholl and Leslie (1999, p. 133): "The essence of architectural modularity is a set of restrictions on informational flow". The sole access that other parts of the mind have to modules is via their inputs and outputs.

8. Modules show developmental trajectories that may be specific to each particular module. Put differently, each developing module appears to be innately channelled to some significant degree.[13]

As indicated, Fodor maintains that the collection of properties outlined above be ascribed primarily to input and output systems at the periphery of the mind. Those systems comprise what we can call *perceptual modules*. Perceptual modules are capable of processing only perceptual information that is given to them directly by the senses. Central cognition, on the other hand, involves *conceptual abilities*, which are decidedly non-modular according to Fodor. Fixing beliefs or making decisions apparently requires one to bring together information from various and importantly different domains. For instance, before I decide to go to the movies, I need to consider whether the movie I want to see is playing, whether I have enough money, whether I would not do better to spend the money on food, whether my friend is joining me, whether the employees at the movie theater are on strike because of an unfavorable economic situation, etc. But if central cognition is capable of integrating information across domains, it cannot itself possess a modular nature. For Fodor, central cognition can fulfill its characteristic function only because central cognition is not isolated in any of the forms characteristic of modules.

(B) Conceptual Modules

Subsequent research in cognitive psychology has nevertheless been motivated to search for modules in addition to the perceptual ones. Its overall agenda has been to uncover *conceptual modules* that support the mind in its central processes of belief-fixation and decision-making. In contrast to Fodor's (1983) perceptual modules, concep-

13 For a discussion of innately channelled development, see Botterill and Carruthers (1999, ch. 3).

tual modules will not be restricted to perceptual information alone. Some conceptual modules may additionally or even entirely process suitably conceptualised output from either perceptual modules or other conceptual modules interacting with them (Botterill and Carruthers, 1999, ch. 3; Sperber, 1994). There are, in fact, data that support the postulation of conceptual modules. Alan Leslie (1987, 1994, 1995) and Simon Baron-Cohen (1995), for instance, argue for a *theory of mind module,* which underlies our everyday practice of attributing mental states to other people in order to explain and predict their actions. In another instance, Scott Atran (1990), argues for a *folk biology module* that enables us to categorise natural kinds via underlying essences that cut across perceptual similarities of shape. Such essentialism in our reasoning about nature is unique, Atran argues, and does not show up in other non-biological cognitive domains.

These and similar kinds of modules are characterised as conceptual because the input they receive and process is already conceptualised in some form that goes beyond simple perceptual information given to the senses. Scientists in favour of conceptual modules agree that many of the Fodorian properties of modules must be relaxed somewhat to accommodate them, but they believe that conceptual modules will nonetheless show sufficiently interesting analogies with perceptual modules to include them as genuine modular cognitive systems (Baron-Cohen, 1995; Botterill and Carruthers, 1999; Sperber, 1994).

While the distinction between perceptual and conceptual modules allows us to organise some of the views about modularity, it is by no means enough to capture all the varieties of modules written about in the relevant literature (Samuels, 1998; Scholl and Leslie, 1999; Segal, 1996). These further varieties cut across the perceptual/conceptual divide and they too need to be outlined to give a complete understanding of the position of evolutionary psychology. One way to characterise modularity that cuts across the perceptual/conceptual divide is proposed by Gabriel Segal (1996). Segal argues that there are two broad types of modules that need to be distinguished: *synchronic* modules (C) and *diachronic* modules (D). I shall discuss these types of modules in turn.

(C) Synchronic Modules

Synchronic modules are cognitive devices that psychologists postulate in order to explain a specific capacity at a given time. The category of synchronic modules, in turn, can be divided further into (modular) bodies of knowledge and (modular) computational mechanisms (Samuels, 1998, 2000; Segal, 1996). I shall call them respectively, *knowledge modules* and *computational modules*.

Knowledge Modules

Cognitive scientists postulate knowledge modules whenever behaviour within a domain is explained by positing a body of knowledge that is *specific* to this domain and that does not apply outside of the domain: knowledge modules are in that sense domain-specific devices.

Domain-specific modules are contrasted with domain-general cognitive devices. Domain-general devices are devices that are capable of solving problems *across a wide range of importantly different domains*. This definition depends on how wide the range of domains must be and on what it means for domains to be importantly different from each other. I will have to explicate this definition in detail because the existence or non-existence of general-purpose cognitive devices plays an important part in evolutionary psychology.

I will tackle the second point of the definition first. If the attribute of "importantly different" is to do any work, it must add something crucial that is missing from the simple reference to "different" domains. The need for "importantly different" domains in the definition of domain-generality is brought out by considering whether cognitive devices can be applied to different domains without thereby automatically counting as domain-general. Dan Sperber (1994) has given arguments for just such a scenario, arguments which depend on the nature of modules and which I will now present here to clarify the question above.

Modules are activated by information only if the information satisfies specific input conditions determined by the module's design and its various interactions with other parts of the mind. Any information that does not satisfy these input conditions consequently will not activate the module. On the basis of these observations, Sperber introduces the following two domains: the *actual domain* and the *proper domain* of a module. As Sperber (1994, p. 51) notes:

> The *actual domain* of a conceptual module is all the information in the organism's environment that may [...] satisfy the module's input conditions. Its *proper domain* is all the information that it is the module's biological function to process.

An example will clarify this distinction. Let us assume that our mind possesses a module that is specialised in computing the emotional states speakers express through the quality of their utterances. Among its most important input conditions we find the rhythm and prosody characteristic of auditory noise. Once activated by utterances that satisfy the input conditions, the module decodes the auditory noise into the emotional states of speakers. Now there are good reasons to think that this information (utterances) constitutes the *proper domain* of the given module, because the module was specifically *selected for* by natural selection to perform this kind of work. However, the organism's environment may contain different types of information that also satisfy the input conditions of the module. For instance, it may be the case that music also displays the right kind of rhythm and prosody, which will consequently lead the module to decode music into emotional states. While the module was not specifically *selected for* the decoding of music (it is not its proper domain), music may still activate the module in the current environment because it satisfies the right input conditions. We can then say that music belongs to the actual domain of the module; it belongs to the set of all information that is actually able to activate the module – regardless of whether or not that information is the biologically proper information that originally led to the development of the module. The conclusion that Sperber wishes to draw from his distinction is that whilst a module may be applied across different domains, this fact alone does not prove conclusively that a cognitive device must be domain-general.

Returning to our problem of defining domain-generality, we can now see what the attribute of "importantly different" domains must add to the definition. For a cognitive device to be properly domain-general it is not enough that it is used across different domains, we must also ensure that these different domains do not belong to the actual domains of one and the same particular cognitive device. That is exactly why our earlier ability of decoding auditory quality and music into emotional states should not be classified as properly domain-general with respect to these abilities. Even though it is applied in such different domains as linguistics and music, the two domains share the *same* input-conditions of one module and so our decoding ability is domain-specific. The only difference between the two domains is that linguistics belongs to the proper domain of the module, while music belongs to the module's actual domain. An essential requirement for a properly domain-general cognitive device, therefore, is *that its respective domains do not share the same input conditions.* A possible example of such a domain-general cognitive device would be one whose biological function is to process information that stems from two (or more) proper domains of different modules. The requirement that the information has to stem from different proper domains ensures that the information does not share the same input conditions of one module. This, again, excludes the domain-specific examples to which Sperber draws attention to in his reasoning. The biological function of the suggested domain-general devices could be to integrate such information with different input-conditions by weighing the information or by prioritising some information over other information.

We are still left with the requirement that a domain-general device must be used in a *wide range* of importantly different domains.

The question now is how wide the range needs to be for the device to count as properly domain-general. Are devices that are recruited in only two domains sufficiently domain-general, or need they be recruited in three or more domains, or even in potentially all different domains? But this requirement is easily dealt with. I will call a cognitive device domain-general if it is actually or potentially applied to at least two *importantly* different domains. We can summarise the discussion of domain-specificity and domain-generality as follows:

(DS) A cognitive device is domain-specific just in case
(a) it is a body of knowledge,
(b) it is devoted to exactly one proper domain,
(c) it does not apply across *importantly* different domains.

(DG) A cognitive device is domain-general just in case
(a) it is a body of knowledge,
(b) it applies to at least two different domains as its proper domain,
(c) the domains are *importantly* different.

Knowledge modules are thus essentially characterised by their use of bodies of domain-specific knowledge in the sense of DS. Furthermore, I shall reserve the notions of domain-specificity and domain-generality to exclusively refer to bodies of knowledge, and not to computational mechanisms manipulating those bodies of knowledge. I therefore need to turn to those computational mechanisms themselves.

Computational Modules

Cognitive psychologists postulate computational modules whenever behaviour in one domain is explained by positing computations that are both *special* to this behaviour and that differ across importantly different domains. Computational modules are therefore sets of computational mechanisms that manipulate bodies of knowledge, rather than the bodies of domain-specific knowledge themselves (Fodor, 1983, 2000).

To distinguish these task-specific computational mechanisms from bodies of domain-specific knowledge, I shall refer to them as *special-purpose computational mechanisms.* Special-purpose computational mechanisms are mechanisms, which are specialised for solving a particular kind of problem, and which are ineffective when used outside of their specific domain. Once special-purpose computational mechanisms are defined, we can contrast them with their opposite: *general-purpose computational mechanisms.* General-purpose computational mechanisms are mechanisms that are capable of being used across a wide range of importantly different domains. Keeping in mind the observations about importantly different domains, we can again summarise the features of special-purpose and general-purpose mechanisms in the following way:

(SP) A cognitive device is special-purpose just in case
(a) it is a computational mechanism,
(b) it is devoted to exactly one proper domain,
(c) it does not apply across *importantly* different domains.

(GP) A cognitive device is general-purpose just in case
(a) it is a computational mechanism,
(b) it applies to at least two different domains as its proper domain,
(c) the domains are *importantly* different.

Computational modules are thus essentially characterised by their use of special-purpose computational mechanisms in the sense of SP.

The distinctions I have just introduced complicate the modularity thesis considerably.[14] It will be thus helpful to place all of the

14 The thesis of modularity is even more complicated than I suggest in my discussion. For my purposes I have concentrated on different notions of cognitive modules. But disciplines such as neurology, developmental sciences,

contrasting features in a matrix and then to systemise the possible versions of the modularity theses of the mind.

		Computational Modules	
		General-Purpose	Special-Purpose
Bodies of Knowledge	Domain-General	Massive Universality	Expert Processor
	Domain-Specific	Expert Knower	Massive Modularity

Fig. 1: Possible Cognitive Architectures[15]

The matrix situates the possible cognitive architectures within the slots provided. At the two extremes lie completely general cognitive architectures that use only general-purpose mechanisms in conjunction with bodies of domain-general knowledge (Massive Universality), and completely specific cognitive architectures that use only special-purpose mechanisms in conjunction with bodies of domain-specific knowledge (Massive Modularity). In between lie cognitive

and genetics employ different versions of non-cognitive modules for their purposes. The following is just a selection:

Neuro-physiological modules: functional subparts of the brain that can be explained in terms of brute neuro-physiological machinery.

Developmental modules: parts of the developing embryo that can develop largely independent of the context in which they occur (Wagner, Mezey, and Calabretta, 2001).

Genetic modules: semi-independent genetic complexes with tightly integrated geneactions and products (Wagner, 1995a, 1995b).

However, these and other forms of non-cognitive modularity are not directly relevant to my discussion of evolutionary psychology, which is a hypothesis about cognition.

15 The matrix is understood to include both perceptual and conceptual modules.

architectures that combine different types of modular structures with different types of non-modular ones (Expert Knowers and Expert Processors).

But where do evolutionary psychologists situate themselves in the matrix above? They tend to gravitate towards the Massive Modularity architecture of the mind. Evolutionary psychologists view the mind as composed largely or even entirely of modules containing bodies of domain-specific knowledge (knowledge modules) and special-purpose computational mechanisms (computational modules). But this definition of evolutionary psychologists' position is not precise enough. For an important distinction can still be drawn between two very different versions of evolutionary psychology – depending on how dense the mind is supposed to be populated by modules. We can distinguish between a strong and a weak version of evolutionary psychology in the following terms:

1. Strong Versions of Evolutionary Psychology:
The mind is composed entirely of knowledge modules as well as computational modules.

2. Weak Versions of Evolutionary Psychology:
The mind is composed largely of either knowledge modules or computational modules.

Strong evolutionary psychology claims that the mind is brimming with specialised cognitive equipment of both variety and that, furthermore, the mind does not use general cognitive capacities to do any important work. Weak evolutionary psychology by contrast does not endorse such strong claims. It only claims that the mind contains large amounts of specialised equipment: the equipment can be of either variety, and the mind does *not* have to possess both knowledge modules as well as computational modules – it is a non-exclusive either or condition. Furthermore, the mind is *not* prevented from using general cognitive equipment that does important work in the mind's overall cognitive economy – given that the general cognitive equipment is embedded in the modular machinery. The critical thrust of weak evolutionary psychology is directed at traditional cognitive sciences that have largely overlooked the predominance of specialised cognitive devices within the mind's structure. To put it differently, weak evolutionary psychology leaves room for general cognitive capacities as long as psychologists also value the fact that our minds will contain massive amounts of specialised equipment alongside whatever additional general capacities the mind might have.

In this work I will adopt the strategy of pursuing the weak version of evolutionary psychology. The reasoning behind my strategy here is to present a version of evolutionary psychology that is as acceptable as possible to get the debate going and to investigate how far the weak version of evolutionary psychology can carry us in our understanding of the evolution and the structure of our mind. An argument for weak versions of evolutionary psychology will be given in the next chapter. For the time being, I wish the reader to assume that weak versions of evolutionary psychology are the proper version to keep in mind.

Despite differences over strength, both versions of evolutionary psychology agree on one fundamental point: from the viewpoint of evolutionary psychology general-purpose cognitive capacities look decidedly suspicious. General cognitive capacities carry the burden of proof, and their postulation needs justification. It is this suspicion about general-purpose cognitive capacities that radically sets evolutionary psychology apart from the rest of the disciplines within psychology. Evolutionary psychologists officially name their view the *Massive Modularity Thesis of the Mind* (Cosmides and Tooby, 1994b; Sperber, 1994) to emphasise their belief in the highly specialised nature of the mind and to contrast it as sharply as possible with models that allow modularity at most only a marginal position in the mind's structure (Fodor, 1983, 2000).

Evolutionary psychology's pet enemies tend to be models that posit cognitive architectures with Massive Universality. These architectures employ neither bodies of domain-specific knowledge nor special-purpose mechanisms.[16] Some evidence that Massive Universality is indeed what evolutionary psychologists have in mind as a contrasting model is the claim, for instance, that "a domain-general evolved architecture is defined by what it lacks: it lacks any content, *either in the form of domain-specific knowledge or domain-specific procedures* that can guide it towards the solution of an adaptive problem" (Cosmides and Tooby, 1994b, p. 94, emphasis added).

As the matrix shows, further possible cognitive architectures involve Expert Knowers. One example of such a cognitive architecture is Richard Samuels' Library Model of Cognition (Samuels, 1998, 2000).[17] Very broadly, the library model of cognition maintains that all the computational mechanisms that the mind uses to fulfil its central processes such as belief-fixation and decision-making are by nature general-purpose: these mechanisms are therefore not supported by special-purpose computational machinery.[18] The mind can nevertheless benefit from specialised cognitive content because it can access different bodies of domain-specific knowledge (knowledge modules) in the process of reasoning about problems in its environment.

Final possible cognitive architectures involve Expert Processors. Such architectures would use special-purpose computational mechanisms that had access only to domain-general knowledge. To my knowledge, there is no worked out model of this kind in the literature, and it is debatable if such a cognitive architecture would be genuinely possible. However, I will not pursue this issue any further here. Now that we have a clearer picture of what synchronic modules are, I turn to diachronic modules.

16 Evolutionary psychologists call these types of theories about cognitive architecture the Standard Social Science Theory of mind. For them, the standard social science theory of the mind is prevalent in all of the human sciences. For a detailed discussion of what they take the standard social science theory of mind to be see Barkow, Cosmides, and Tooby, (1992).

17 I will discuss Samuels' model in more detail in chapter II.

18 However, he does allow the existence of modular percepetuo-motor input/output mechanisms working at the periphery of the mind.

(D) Diachronic Modules

Diachronic modules are cognitive devices that psychologists postulate in order to explain the *development* of particular cognitive capacities over time. Such modules are thus concerned directly with issues of development and can thus answer many criticisms levelled against the role of modules in cognitive science. For example, one prominent criticism of modules states that they are inappropriate to account for the development of cognition (Gopnik and Meltzoff, 1997; Gopnik and Wellman, 1997; Karmiloff-Smith, 1992, 2000). A capacity such as a theory of mind obviously undergoes development (Baron-Cohen, 1995; Wellman, 1990), whereby the performance of the capacity apparently gets more and more sophisticated as a result of learning. The worry many researchers have about modules is whether such static structures can explain the dynamic processes a developing competence involves. Mostly, such worries are nurtured by a view that sees modules as existing full-blown and fully specified in the mind of infants (or even neonates) (Elman, Bates, Johnson, Karmiloff-Smith, Parisi, and Plunkett, 1996).

But such a view is misguided. As Scholl and Leslie (1999, p. 136) rightly emphasise "there is nothing in the notion of modularity which prevents even 'matured' modules from developing". As we have seen, modules are characterised, among other things, by informational isolation, which means that the information flow between modules is restricted in some interesting way. But there is nothing that would prevent the processes *inside* individual modules from developing. While it is true that restrictions on the flow of information limit the amount of information that is available to developmental processes within a specific module, in and of itself informational isolation does not make internal modular development impossible.

It is the characteristic cognitive function of diachronic modules that they direct or channel the development of cognition in genetically determined patterns of growth. One way of clarifying the concept of diachronic modules is to see them as devices that reliably produce mature cognitive capacities when suitably prompted by environmental interactions. They are a kind of black box that takes experience as input, and produces cognitive mechanisms or bodies of knowledge as output (Segal, 1996). This transition from input to output is (most likely) mediated by sets of innate rules and principles that channel the development in the correct direction.

After so many distinctions it is helpful to briefly restate the main points of this subsection. Evolutionary psychologists are into massive modularity, where modules are construed as functional subparts of the mind that work in relative isolation from each other and that are highly specialised to perform their job. As has been shown, the term "module" in fact applies to a very broad range of cognitive structures: modules can be either perceptual or conceptual, they can be either bodies of knowledge or computational mechanisms manipulating such knowledge, or modules can be either synchronic or diachronic. So when evolutionary psychologists talk about modules, the whole variety of modules must be kept in mind. I also stated that I am going to pursue a weak version of evolutionary psychology from now on. This version of evolutionary psychology is weak because it only claims that the mind contains a large amount of specialised cognitive structures of knowledge modules or computational modules. This constraint by itself leaves room for the postulation of general cognitive structures as long as these cognitive structures occur side by side with large amounts of specialised ones.

1.3 Cognition and Cognitive Adaptation

The concepts of *computation* and *modularity* belong to the common toolbox of cognitive science, and are not special to evolutionary psychology. What is special about evolutionary psychology is its additional belief that modules are *biological adaptations*. According to this assumption, the mind is formed and shaped by selective forces that operated on it over millions of years to result in a system of specialised mental machinery for solving problems that (ultimately) affected the reproductive success of our ancestors. In introducing the concept of adaptation into their theorising, evolutionary psychologists take on a complex notion that is well entrenched and regimented within evolutionary biology and its sub-disciplines (Park, 1996; Williams, 1966). I shall highlight those aspects of adaptations that are most relevant for an understanding of the project of evolutionary psychology.

When biologists talk about adaptations, for example, wings or beak sizes, they typically refer to phenotypic traits[19] that organisms possess because these traits have given their ancestors increased reproductive success over competing con-specifics. By systematically and incessantly favouring those varieties of a given phenotypic trait that work better in solving (relevant) problems, natural selection is able to engineer an ever tighter fit between the structure of the traits and the particular environment exploited by them.[20] Whilst it is true that natural selection is not the only process that causes evolution in a population,[21] it is the only known process that can produce adaptations.

Environment of Evolutionary Adaptedness

Adaptation is a relative notion: adaptations are always adaptations for solving particular problems. They therefore need to be analysed

19 A phenotype is a complete description of all the visible properties of an organism such as molecular, neuro-physiological, anatomical, behavioural, and mental properties.

20 An influential tradition in biology treats genes as the real units of natural selection, because it is genes that help building phenotypic features of an organism, and it is genes that get transmitted to the next generation (Dawkins, 1989; Williams, 1966). There are now interesting models of how natural selection acts upon the genetic material responsible for the development of cognitive modules. For such a discussion see especially Wagner (1995a, 1995b), Wagner and Altenberg (1996).

21 Typically evolution is analysed in evolutionary genetics as changes in gene frequency within a natural population. Apart from natural selection, changes in gene frequency can be caused by genetic drift, neutral selection, etc. (Ayala, 1982).

in relation to these problems. Not just any problems, though, but specific problems. The relevant problems are recurrent problems ancestors repeatedly faced in their environment and had to solve to survive or reproduce. Such problems are known in the literature as *adaptive problems*, and a complete listing of all the adaptive problems a particular species faces during its evolution constitutes the species' *environment of evolutionary adaptedness*. There are two features of the environment of evolutionary adaptedness that need special emphasis.

Firstly, the environment of evolutionary adaptedness is *not* a specific time or geographic place. Rather, it is a conglomeration of adaptive problems that need not have arisen at the same time or place. Some cognitive tools of organisms might be an old heritage they share with other species, other cognitive tools might be new acquisitions that are unique to the species. Humans, for example, have face recognition abilities that might be as old as 20 Mio years and which they share with the whole order of primates, other abilities such as language might be the most recent acquisition (100'000–250'000 years), and one that is unique to humans. The important point is that the adaptive problems that caused the development of different cognitive tools might have arisen at different periods and places in the species' evolution.

Secondly, the definition of the environment of evolutionary adaptedness hints at a distinction one should draw between ancestral and modern environments. The environment of evolutionary adaptedness is the *past* environment to which the ancestors of a species had to adapt, and in relation to which they developed their array of cognitive adaptations. But if there is a mismatch between the environment of evolutionary adaptedness and modern environments, the cognitive adaptations that have worked well in the former might not do so equally well in the latter. In past environments, for example, it was clearly adaptive for our ancestors to gather sweet substances, because they were both nutritionally valuable and very sparse. Today, however, in our modern environments with fast-food chains our cravings for sweet substances may easily lead to grave health-problems. Our inherited biases cannot cope with those relatively recent environments. The behaviour that these biases cause will turn out to be maladaptive in modern environments.

Evolutionary psychologists strongly emphasise the possibility of mismatches between past and modern environments because they believe that such mismatches support their claims that the concept of adaptation applies primarily to the cognitive structures causing behaviour, and not to the behaviour itself (Cosmides and Tooby, 1994a; Symons 1987, 1989, 1990, 1992; Tooby and Cosmides, 1990). Some theorists, on the contrary, wish to apply the evolutionary consideration to the behaviour of organisms directly, literally counting babies to assess the reproductive fitness of organisms (Betzig, 1997). However, evolutionary psychologists believe that such intention is misguided. As indicated, a possible mismatch between past and modern environments can seriously disrupt the normal functioning of cognitive adaptations, and consequently distort whatever data scientists collect on the basis of actual behaviour.[22] Such distortions of the behavioural data can have several causes. First, the case may occur that unsuccessful behaviour in current environments is still caused by cognitive adaptations that

22 Darwinian anthropologists (Alexander, 1974, 1979, 1990; Betzig, 1997; Crawford and Krebs, 1998; Irons, 1979, 1990; Turke, 1990a, 1990b) are especially prone to such mistakes, according to evolutionary psychologists. Darwinian anthropologists assume that it is unnecessary to invoke (adapted) special-purpose cognitive capacities when studying behaviour from an evolutionary perspective. Some of them postulate general-purpose mechanisms that compute fitness costs and benefits and adjust behaviour adaptively to whatever environment the individual finds itself in. The only adaptation that the human mind possesses, then, is a general (inclusive) fitness-maximizer that causes human behaviour to assume the form that best maximises fitness in the current environment. According to Darwinian Anthropologists, therefore, scientists should expect behaviour to be adaptive in contemporary environments too. Their focus is more on the actual behaviour. Evolutionary psychologists, on the other hand, caution against such expectations. They wish to highlight the fact that we should only expect our adaptations to generate adaptive behaviour in the ancestral environment or in modern environments that are relevantly similar to the environment of evolutionary adaptedness.

worked well in past environments. It just happens that features of modern environments disrupt these cognitive adaptations, but these adaptations still belong to the furniture of the mind. Second, the case may occur that successful behaviour in modern environments is not caused by cognitive adaptations at all. Take for example, the case of writing. It is simply not possible that humans could have developed a cognitive adaptation to write, simply because the origin of writing is too recent an invention for natural selection to build an adaptation for it.

In both cases the inference from current behaviour to cognitive adaptation is not justified. This is not to say that evolutionary psychologists completely dismiss data stemming from actual behaviour. But they caution against it, and include it in their theorising only if empirical studies show that modern environments are relevantly similar to the environment of evolutionary adaptedness of the species studied.

Identification of Adaptations

Not all traits of organisms are adaptations, and the detection of an adaptation is a major breakthrough in the evolutionary understanding of a given species. Adaptations, however, are not always easily detectable. Scientists need a notion of reliable indicators of adaptive design that can guide them in their search. Commonly, three such indicators are suggested: *design complexity*, *efficiency in solving problems*, and *constancy in development* (Williams, 1966). The first indicator is exemplified by a design that exhibits many subparts that are integrated in non-random ways to carry out a vital function. A classic adaptation adhering to this principle is the exquisitely complex human eye. Many subparts of the eye, e.g., light cones and lenses, work together to make vision possible. The second indicator is exemplified by the *efficiency* with which adaptations solve their problems (taking into account the energy costs the maintenance of such a structure demands). The third indicator is exemplified by the *constancy* with which adaptations emerge during development. Structures that reliably develop in populations are likely to be adaptations that were specifically designed to carry out a vital function for the organism (Williams, 1966).

Adaptations, moreover, are structures that originated by natural selection because of their positive influence on the organism's reproductive success. But beneficial behaviour alone is never sufficient to infer the occurrence of adaptations. For some beneficial behaviour may be due to cognitive structures that did not originate by natural selection for their *current* effects on the organism; it might thus not be the case that the current effects had any role to play in their original development. Such cognitive structures are known in the literature as *exaptations*. There are two forms of exaptations: *pre-adaptations* and *spandrels*.

Pre-adaptations are traits that originated because of a particular purpose they carried out for an organism, but which were later co-opted for another purpose instead (Gould and Vrba, 1982). Bird wings, for instance, are assumed to have been originally adapted for the purposes of thermoregulation in reptiles, but later became co-opted for flying and are now currently benefiting birds in the way they do. But even though wings support flying in birds today, they did not *originate* by natural selection *for* their current function of flying (even though natural selection now maintains wings in the population of birds because of their beneficial effects for flying).

Spandrels, in contrast, are traits that originally had no adaptive consequences for the organism whatsoever. They may, for example, have originated merely as developmental by-products of other traits that had been selected for, but were later co-opted to carry out an adaptive purpose on their own (Gould and Lewontin, 1979). Again, the structure did not arise originally through natural selection for its current adaptive purpose.

Strictly speaking, then, exaptations are *not* adaptations because they did not originate through natural selection for their current effects on the organism's reproductive success. Evolutionary psychologists therefore need to rule out exaptations before they can postulate adaptations underlying specific capacities. Now that does not mean that evolutionary psychologists should not be interested in exaptations, they clearly should be so interested if they want to further the evolutionary understanding of a particular species. But as long as they claim that modules are biological adaptations, they will have to distinguish these modules from other structures of the organism that are exaptations.

As has been shown in this subsection, treating modules as adaptations carries many commitments evolutionary psychologists have to take on board. Among them is a good knowledge of the environment for which these adaptations evolved and a good understanding of the properties that distinguishes adaptations from exaptations.

1.4 Evolutionary theory's methodological use

Evolutionary psychologists use their knowledge of the evolutionary process as a methodological tool to derive informed guesses about the structure and organisation of the human mind. Their belief is that once cognitive scientists start to develop hypotheses based on insights from evolutionary biology, these scientists will be able to devise better experiments that make possible the discovery of information-processing mechanisms that are missed by other methodologies.

What are the assumptions that fuel those hopes? Evolutionary psychologists believe that the tight fit that exists between an evolved phenotypic trait and the adaptive problem it solves is enough of a guarantee to use theories of adaptive problems as a theoretical tool. In a typical scenario, evolutionary psychologists will look at what is known about our ancestral environment and use whatever knowledge they can acquire about the problems our ancestors would most likely have had to solve to derive hypotheses about the most likely design features cognitive capacities would have to have to solve such problems efficiently.[23]

23 Examples of the application of such methodology to several problem-domains can be found in: Cosmides, 1989; Cosmides and Tooby (1987, 1994b, 2000), Tooby and Cosmides (1990, 1992).

Since the process of adapting an organism to its environment has two aspects – the adaptation (product) and the adaptive problem (cause) – the methodology of evolutionary psychology can be conducted in two different ways. These different ways give rise to two distinct research projects within evolutionary psychology.

The explanatory project

The explanatory project starts by documenting particular cognitive devices that are independently known to exist in the mind. Thus, given a known operative psychological mechanism, evolutionary psychologists aim to understand the underlying evolutionary reasons *why* such a mechanism evolved in the first place. In this kind of research, they need to:

> start with a known phenotypic phenomenon, such as pregnancy sickness, language, or color vision, and try to understand what its adaptive function was – why that design was selected for rather than alternative ones. To do this, one must show that it is well designed for solving a specific adaptive problem, and that it is not more parsimoniously explained as a by-product of a design that evolved to solve some other problem. (Barkow, Cosmides, and Tooby, 1992, p. 9)

In other words, investigation begins with the end product of the evolutionary process, i.e. the cognitive adaptation, and works backwards to the historical cause responsible for its origin and current persistence in the population. But why should psychologists bother to find out about the history of a mechanism if the mechanism is already known to exist? Evolutionary psychologists argue that this type of research ultimately helps scientists to understand better the intricate details of the computational machinery. The knowledge of the exact selection pressures to which the mechanism responded can act as a source of new insights into the detailed workings of known mechanisms.[24]

24 This kind of reasoning is common in engineering and is there called "reverse-engineering" (Pinker, 1996): an engineer finds herself with a complex structure whose function she does not yet understand. The goal of reverse-engineering is then to find out what the equipment actually does and how it does it.

The predicative project

The predictive project, by contrast, starts with the adaptive problems and predicts the existence and structure of currently unknown cognitive adaptations that are able to solve the adaptive problem:

> The approach advocated here is the reverse: first, one uses existing and validated theories from evolutionary biology to define an adaptive problem that the human mind must be able to solve, and to deduce what properties a psychological mechanism capable of solving that problem must have. Then one tests to see whether there is evidence for a psychological mechanism having the hypothesized properties. It is a constrained and predictive approach, rather then a compilation of post hoc explanations for known phenomena. (Cosmides and Tooby, 1989, p. 190)

Evolutionary biology, Cosmides and Tooby believe, can help researchers to home in on the adaptive problems our ancestors most likely encountered in their environment of evolutionary adaptedness. Accordingly, then, a detailed task-analysis of those problems together with knowledge derived from hunter-gatherer studies and comparative studies of primate social societies will give researchers valuable clues to enable them to detect and map previously unknown computational structures within the mind whose only raison d'être is to solve those problems. Such research has a threefold structure, which evolutionary psychologists have to follow:

(1) Develop hypotheses about which adaptive problems our ancestors most likely faced in their ancestral environments.

(2) Based on the outcome of (1), derive hypotheses about the required computational design features needed to solve those problems. The thorough and detailed analysis of the task to be solved will tightly constrain the designs that can solve it, and will thereby allow the researchers to exclude *a priori* designs that could not solve the task.

(3) Based on the results of (2), devise experiments to pick out the solution the organism actually adopted to solve the adaptive problem. This final step is necessary because even a detailed analysis of a given task will never uniquely determine all the minute details of the computational structure. There will always be alternative solutions to a particular problem, and experiments are needed to decide among them.

It will help later discussion to visualise these two research projects thus:

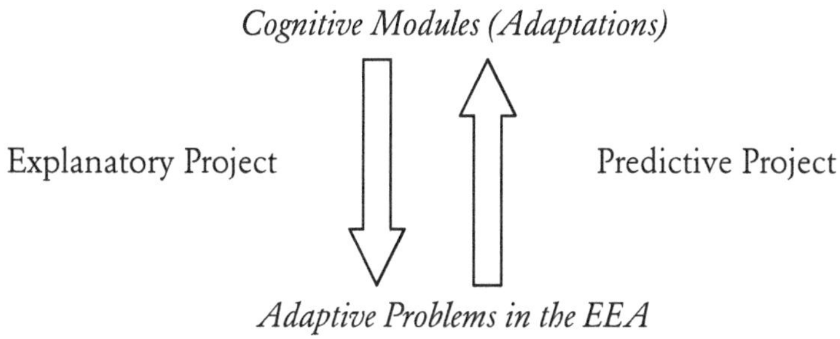

This concludes my discussion of the core concepts of evolutionary psychology. As I have shown, two of these assumptions, computation and modularity, are not special to evolutionary psychology.[25] The specific assumptions that sets evolutionary psychology apart from the other cognitive sciences is their reliance on adaptation, massive modularity and their methodology to derive seemingly fruitful hypotheses about the structure of the mind. But how are these views supported? It is to this question that I shall now turn.

25 This does not mean that all of cognitive science endorses these concepts unanimously. There is still considerable debate about the applicability of computation and modularity in the study of the mind (Elman, Bates, Johnson, Karmiloff-Smith, Parisi, and Plunkett, 1996; Fodor and McLaughlin, 1990; Fodor and Pylyshyn, 1988).

2. Theoretical arguments of evolutionary psychology

I have introduced the model of the mind that evolutionary psychologists advocate – the evolved massively modular mind. It is now time to present the theoretical arguments evolutionary psychologists use to support their controversial model. While many scientists currently investigate and sample data about specialised cognitive capacities,[26] the direct application of the evolutionary consideration to the cognitive domain is still a minority position. Many scientists do not believe that applying the theory of natural selection to the mind's functioning can give us any deep insights into the nuts and bolts of our cognitive architecture, much less its massively modular structure. It is therefore important to outline the theoretical arguments evolutionary psychologists use to support their central claims. As we shall see, most evolutionary psychologists' arguments are aimed at showing that general-purpose mechanisms are suspicious from an evolutionary point of view. So their intention is to establish massive modularity by both undermining its competitor, and simultaneously proving that modules are the more plausible cognitive structures from an evolutionary point of view.

2.1 The Argument from Methodology

The first argument evolutionary psychologists use to support their view is of a methodological nature. The argument is not so much an argument *about* a particular cognitive structure as it is an argument *for* a particular way of deriving hypotheses about a cognitive structure. I explain what I mean by this.

26 For such research see: Chomsky (1988), Gazzaniga and Altman (1998), Gopnik and Meltzoff (1997), Hirschfeld and Gelman (1994), and Karmiloff-Smith (1992).

The aim of cognitive scientists is to construct a complete list of all the functional subparts that the mind uses in its reasoning. Because a lot of the cognitive tasks the mind performs will not, as I already explained in the introduction, be easily accessible to introspection, it is imperative that cognitive scientists possess reliable principles that can guide them safely through the cognitive architecture. Cognitive scientists, therefore, need a methodology in the light of which they are able to devise the right experiments to reveal the mind's underlying cognitive structure. Without an effective methodology the progress of cognitive science will be seriously jeopardized, for it will be bogged down by the sheer number of possible cognitive design features that the mind may instantiate.

Consequently, every branch of cognitive science uses its own methodology to inform experiments aimed at uncovering the *what* and *how* of cognitive devices. Generally, those methodologies are underwritten by a particular understanding of the concept of a trait's function. Once you know *what* a device does, once you know the function of the device, you can start to figure out *how* it does its function. An understanding of function is thus already presupposed in cognitive science. Evolutionary psychologists, however, believe that

> most cognitive scientists proceed without any clear notion of what 'function' means for biological structures like the brain, or what the explicit analysis of function could teach them. (Cosmides and Tooby, 1994a, p. 43)

Note that evolutionary psychologists are not saying that cognitive scientists have a *confused* notion of function. Rather, they are saying that cognitive scientists have only an *implicit understanding* of function that is mainly informed by intuitive knowledge of how objects work. Evolutionary psychologists believe that as a consequence of such implicit and intuitive understanding cognitive scientists have hitherto been interested exclusively in developing computational models for cognitive capacities that are easily accessible by the intuitive knowledge of function – such as perception, memory, and general problem solving. Evolutionary psychologists believe that an *explicit understanding* of function based on evolutionary theory will help cognitive scientists to uncover cognitive devices that were not previously accessible by other methodologies.

The argument from methodology can be made explicit as follows:

(P1) Psychological theorising needs an understanding of function. Knowing what a structure *is for* will help us to better understand how it does what it does.
(P2) Evolutionary biology offers an explicit and theoretically motivated understanding of function that fulfils the requirement of premise (P1).
(C) Therefore, psychology should turn to an evolutionary concept of function.

Evolutionary psychologists firmly believe that once cognitive scientists start taking the evolutionary understanding of function more seriously, progress in charting the mind's cognitive equipment will follow suit. The argument from methodology alone, however, is not sufficient to argue for the massively modular structure of the mind. At best it provides a research strategy, which moves from the (evolutionary) function of a subpart to how that function is performed, once the function of the subpart is uncovered. But to argue for a *massively* modular structure of the mind, this research strategy needs additional support from outside. One such additional support might be the fact that particular species encountered a *massive* number of importantly different adaptive problems that led to the development of *massive* numbers of specialised subparts dealing with those problems. The following theoretical arguments that evolutionary psychologists use are meant to give additional support to the claim about the massively modular mind.

2.2 *The Argument from the Trend to Specialisation*

The second argument in support of the massive modularity depends on the logic of the evolutionary process (Cosmides and Tooby, 1987, 1994a; Tooby and Cosmides, 1990, 1998). This argument claims that the multitude of specialised tools in the realms of the natural, the artificial, and the mental is not an accident, but testifies to the manner in which evolution works in solving adaptive problems. We do not have, for instance, a general organ that is responsible for digestion and locomotion. These functions are discharged by two very different structures, the stomach in the former case and the skeleton in the latter. The trend to specialisation that is manifest in such examples is a crucial element of how the evolutionary process works.

But why should that be? The reasons are straightforward. Just imagine under what conditions a general organ could work. Because it would apply to two or more very different problems, a general organ would need to embody a compromise. To count as a proper general solution to two or more problems, the organ must deny itself any problem-specific elements that would bias it towards the one problem. Giving away elements that are specific to one or the other problem, however, means giving away precious information the organ could use to efficiently solve one of the problems. Consequently, a domain-general organ's overall performance on the two problems would deteriorate. But if the organ tries to incorporate problem-specific information into its design, this information will conflict with the demands of the other problem. It would appear that a better way to deal with each problem would be to construct two or more specialised organs that could each fully exploit the task demands of their respective problem. Moreover, because specialised organs are capable of using specific bodies of knowledge or mechanisms such organs will be fast, reliable, and effective in solving their problems, and in the long run would out compete any domain-general organs along exactly these dimensions. Suitably generalised, the same kind of consideration applies to adaptive problems at large because,

> Different adaptive problems often require different solutions and different solutions can, in most cases, be implemented by different, functionally distinct mechanisms. Speed, reliability and efficiency can be engineered into specialized mechanisms because there is no need to engineer a compromise between different task demands. (Cosmides and Tooby, 1994b, p. 89)

Reflecting on the logic of the evolutionary process, then, a general trend towards specialisation can be observed. Evolutionary psychologists have turned this insight into the following rule,

> As a rule, when two adaptive problems have solutions that are incompatible or simply different, a single solution will be inferior to two specialized solutions. (Cosmides and Tooby, 1994b, p. 89)

Their claim is that the same trend towards specialisation that is manifest in morphological structure is also manifest in cognitive structure. According to evolutionary psychology, general cognitive solutions are as improbable as general organs. As Symons (1992, p. 142) puts it,

> it is no more probable that some sort of general-purpose brain/mind mechanism could solve all the behavioral problems an organism faces (find food, choose a mate, select a habitat, etc.) than it is that some sort of general-purpose organ could perform all physiological functions (pump blood, digest food, nourish an embryo, etc.). There is no such thing as a 'general problem solver' because there is no such thing as a general problem.

Symon's argument here is that adaptive problems[27] that have solutions that are (a) incompatible or (b) simply different will produce

27 It is obviously not true that adaptive problems always give rise to cognitive solutions. Many adaptive problems might give rise to molecular solutions (digestion of particular valuable food), morphological solutions (thick fur against cold temperatures), anatomical solutions (teeth for food ingestion), etc. Cognitive adaptations are the set of solutions that are essentially describable in terms of informational states and computational procedures.

selection for bodies of domain-specific knowledge or special-purpose mechanisms. It is further assumed that such modules would have systematically outperformed bodies of domain-general knowledge or general-purpose mechanisms in the human EEA by being faster, more reliable, and certainly more effective. The winners of such an evolutionary race would thus have been organisms with distinctive modular architectures.

2.3 The Argument from Statistical Information

A third argument that evolutionary psychologists use in support of massive modularity centres on the claim that some information that is crucial in solving adaptive problems cannot be acquired by an organism using solely general-purpose learning mechanisms (Cosmides and Tooby, 1994a; Tooby and Cosmides 1990). Why is this so?

Evolutionary psychologists maintain that the major weakness of general-purpose mechanisms unaided by specialised and dedicated cognitive machinery lies in these mechanisms' blindness to statistical information that emerges only over many generations. But it is just such types of information that are crucial for solving adaptive problems. Note that adaptive problems are types of situations our ancestors repeatedly faced and the solution of which promoted their reproductive success. These types of situations involved contingencies and conditions the "knowledge" of which would have aided our ancestors to solve them efficiently, but which our ancestors could not learn because these contingencies emerge only over many generations. There are many reasons for this inability. The effects on the individual's fitness of the statistical regularity embedded in a situation type may be so small that its phenotypic traits could not detect it. Moreover, general-purpose mechanisms are limited to knowledge that can be validly inferred from the perceptual information the organisms receive in their lifetime. This latter fact is due to the absence of specialised elements that the mechanisms could bring to bear on the interpretation of perceptual information. An example will clarify this problem.

Margie Profet (1992) investigated the phenomenon of pregnancy sickness in humans, and she suggested an interesting adaptive explanation of it. Pregnancy sickness usually begins within 2 to 4 weeks after conception, and falls off after 8 weeks. During pregnancy sickness women display a heightened sensitivity to foods with pungent or bitter odours and tastes such as highly spiced foods and all but the most bland vegetables. As Profet found, the onset and the end of pregnancy sickness coincides precisely with organogenesis, the embryonic period that involves formation of the limbs and all other organ systems. Organogenesis is also the period of maximum vulnerability to particular types of plant toxins, which can lead to malformations or even death depending on the dose. Women's intolerance to highly spiced foods during pregnancy sickness may thus have a reasonable adaptive explanation: the various pungent aromas and bitter tastes of plant foods are reliable indicators of the toxins they contain, and it is these toxins that are harmful to the developing embryo.

Now, it is hard to see how ancestral females could have learned this kind of information from perception alone. Note that most plant toxins are not detrimental to the adult female, because she has digestive systems that can break down the toxins very easily. Female humans have such digestive systems because plants and vegetables played a very important part in our ancestors' diet. So most of the time, there would not have been a great incentive to learn about the toxins anyway. Moreover, sometimes females would not observe the detrimental consequences of ingesting plant toxins during a pregnancy, because some malformed embryos would have spontaneously aborted very early on. Finally, many cases of moderate intake of plant foods would not have had a detrimental effect on the developing embryo anyway. So, from a fitness point of view, the females would not have been able to tell whether they had eaten too much or just the right amount of plants during their pregnancy. All these observations taken together make it particularly hard to understand how individual females could have learned this complex information on the basis of general-purpose mechanisms alone. And in any case determining the causes would be an incredibly difficult task, even for a modern scientist. Most of the knowledge crucial for solving the adaptive problem of pregnancy would have depended on statistical information that emerged only over many generations.

However, whilst the organisms cannot learn the crucial knowledge allowing them to solve adaptive problems, natural selection can. By steadily sifting through small (genetic) modifications of cognitive structures, selecting for those structures that are beneficial to the organism, selecting against those structures that are detrimental to it, natural selection can internalise the relevant statistical information into the design of cognitive structures. Thus evolution will build bodies of domain-specific knowledge or special-purpose mechanisms that support the organism in producing adaptive behaviour in the face of adaptive problems. Moreover, such modules will systematically outperform general-purpose cognition because they can exploit specialised content relevant to the specific problems at hand.

2.4 The Argument from Combinatorial Explosion

A fourth argument that evolutionary psychologists use to support massive modularity centres on computational problems that affect general-purpose devices (Cosmides and Tooby, 1994a, 1994b). Evolutionary psychologists suppose that general-purpose devices have only limited access to types of specialised and dedicated cognitive mechanisms. When such devices are confronted by reasonably complex situations, they have to compute most or even perhaps all alternative solutions to a given problem to find out which solution is the best. In so doing, the devices cannot resort to domain-specific information that tells them which particular scenarios are relevant and which are not (and can thereby be safely disregarded). Without appropriate support, the devices are forced to navigate through the whole space of alternative solutions that their computational ingenuity could come up. As the complexity of a given task increases, so too does the number of alternative solutions the device has to consider. The trouble is that alternative solutions increase exponentially. So unless general-purpose devices possess effective defences against exponential explosion, they will soon be overburdened by the computations and will come to a sudden standstill. But can a general-purpose device build up such effective defence strategies? Cosmides and Tooby (1994a, 1994b) seem to think that it cannot.

Of course there are domain-general ways to limit the search of alternative solutions: one can, for example, arbitrarily decide to stop the search whenever the fourth alternative has been computed. But evolutionary psychologists would probably want to say that such domain-general strategies could hardly be called effective and reliable in finding the most adaptive solution to a given task. According to evolutionary psychology, then, natural selection can be expected to produce cognitive adaptations to escape the costly computations; it is expected to select mechanisms that operate only on the basis of limited information and/or computational mechanisms. Cognitive systems with domain-specific bodies of knowledge and/or general-purpose computational mechanisms are an effective mechanism for narrowing down the relevant slots in the decision-matrix, which face the organism at a particular time.

Evolutionary psychologists conclude that natural selection has built many such cognitive adaptations to circumvent the exponential explosions lurking in the particular tasks the mind needed to solve.

3. Conclusion

Evolutionary psychologists start with the assumption that the human mind was caused by evolution. Although many scientists seemingly accept this much, they are not equally prepared to accept the implication that follows from it. According to evolutionary psychologists, the most important implication of the evolutionary perspective is the resulting massively modular mind. The mind thus contains a pervasive modular structure, including some modules that support perception and others that support "higher" cognitive processes involved in decision-making and belief-fixation. While this conclusion may surprise some cognitive scientists, evolutionary psychologists argue that the correct understanding of evolution should compel them to accept massive modularity. Additionally, evolutionary psychologists believe that once massive modularity has been generally accepted in cognitive science, then progress in the study of mind has started.

In this chapter I introduced evolutionary psychology by discussing concepts of computation, modularity, and adaptation. The discussion of these concepts allowed me to draw important distinctions, which are necessary to understand the details of evolutionary psychology. Prominent among these distinctions was the distinction between strong versions of evolutionary psychology and weak versions of it. While strong versions of evolutionary psychology claim that the mind is composed entirely of knowledge modules and computational modules, weak versions of evolutionary psychology only claim that the mind is composed largely of knowledge modules or computational modules. Additionally, weak versions of evolutionary psychology are able to accord general-purpose cognitive capacities a place within the mind's structure provided that these capacities work in conjunction with the mind's modular machinery and that an evolutionary account of them can be given. I therefore proposed to investigate weak versions of evolutionary psychology to see how far we can go with them in understanding our mind's structure. I finally presented evolutionary psychology's theoretical arguments in support of the mind's massive modularity: namely, the argument from methodology, the argument from specialisation, the argument from statistical information, and finally the argument from combinatorial explosion.

Thus, evolutionary psychology seems to present a promising research agenda for psychologists interested in the mind's structure. We must, however, critically assess the concepts and theoretical arguments of evolutionary psychology before any firm conclusions about its consistency can be drawn. The critical assessment of evolutionary psychology is the purpose of the next chapter.

II. Criticisms of Evolutionary Psychology

In the last chapter we saw that evolutionary psychologists argue for the integration of cognitive psychology with the rest of the evolutionary sciences. According to evolutionary psychologists, the benefits of this integration for cognitive psychology will be large. But the adjustment of cognitive psychology to evolutionary theory will lead to major conceptual and practical changes within the field. It should therefore come as no surprise that the reactions from cognitive psychologists to the emerging field of evolutionary psychology have been very mixed. As a consequence, cognitive psychologists have raised important objections to different aspects of evolutionary psychology. The basic tenets of evolutionary psychology, and particularly its massive modularity thesis, remain a hotly debated issue within the scientific community at large.

The aim of this chapter is to assess these objections and to defend evolutionary psychology against them. I do this by putting a particular version of evolutionary psychology to the test. Recall from the last chapter that there exist two versions of evolutionary psychology: strong versions and weak versions, which we defined as follows:

1. *Strong Versions of Evolutionary Psychology*:
The mind is composed entirely of knowledge modules *and* computational modules.

2. *Weak Versions of Evolutionary Psychology*:
The mind is composed largely of knowledge modules *or* computational modules.

While strong versions of evolutionary psychology dismiss general-purpose cognitive capacities, weak versions of it accord general-purpose cognitive capacities a place within the mind's structure – provided that these cognitive capacities are integrated with the mind's modular machinery and that an evolutionary account of them can be given.

For the purposes of this chapter, the weak versions of evolutionary psychology form the proper topic of debate[28] for they have at least the initial advantage of formulating a substantial view of the mind that is less controversial than the one propounded by strong versions of it. Cognitive psychologists will therefore be more inclined to enter a critical discourse about the validity of such versions of evolutionary psychology; which is exactly what we need initially. The chapter will then supplement further arguments in favour of weak versions of evolutionary psychology. Thus, the subjection of weak versions of evolutionary psychology to the objections will bring to the fore whether they are able to offer a consistent and fruitful empirical programme for future research into the mind.

In this chapter I will argue that while some objections call for small changes in evolutionary psychology, none of them do anything to call into question the basic tenets of it. I proceed as follows: Section 1. first evaluates objections to evolutionary psychology that aim at both the methodology and the core concepts. Section 2. then examines an alternative account of the mind proposed in the literature by Richard Samuels that directly engages with evolutionary psychologists' theoretical argument. Samuels wants to show that these theoretical arguments are too weak to favour some versions of evolutionary psychology over his own alternative account of the mind.

28 As will be demonstrated in section 2, where Richard Samuels' arguments are discussed, the prominent position of weak versions of evolutionary psychology throughout this chapter is supported by strong arguments. I ask the reader to accept the weak versions of evolutionary psychology at the outset and to put off its argumentative basis until we reach section 2.

1. Objections to evolutionary psychology's basic tenets

A good way to organise the arguments against evolutionary psychology is by quickly reminding ourselves of evolutionary psychology's core claims:

(1) Talking about adaptive problems and the evolutionary functions of cognitive capacities solving them is a necessary methodological tool to chart successfully the cognitive architecture.
(2) The cognitive architecture is composed largely of either bodies of domain-specific knowledge *(knowledge modules)* or special-purpose computational mechanisms *(computational modules)* that are adaptations to the evolutionary environment.

Arguments against evolutionary psychology can now be assigned either to the methodology or to the core concepts of evolutionary psychology. I will start my evaluation of the arguments with the claims about methodology.

1.1 Objections to the methodology

(A) Evolutionary function vs. current function

As we saw in the last chapter, cognitive capacities solve adaptive problems for it is their evolutionary function to tackle these problems. Moreover, evolutionary functions explain *why* cognitive capacities originated in the species since evolutionary functions are those things cognitive capacities normally do that had a positive impact on the ancestors' reproductive success. Now the evolutionary function of a trait lies at the very core of evolutionary psychology's methodology. Evolutionary psychologists believe that once scientists accept this methodology they will benefit from two research projects: the explanatory project that moves from the cognitive capacity to its evolutionary function (solving the adaptive problem) and the predictive project that leads from the evolutionary function (solving the adaptive problem) to the cognitive capacity performing it.

Most psychologists will happily agree that an understanding of the function performed by a cognitive capacity is necessary to understand the design features of the cognitive capacity such as its representations and/or computational algorithms. Since at least David Marr (1982), cognitive psychologists have been familiar with inferring hypotheses about design features of cognitive capacities on the basis of the functions performed by them. Marr (1982) developed a general account of information-processing systems in terms of three levels of analysis:

Level 1: Computational theory
This level of analysis characterises a given cognitive capacity in virtue of the function it performs in its interaction with the environment.

Level 2: Algorithms and Representations
This level of analysis lists the types of algorithms and representations the capacity uses to perform its function specified in the computational theory.

Level 3: Implementation
This level of analysis investigates the underlying hardware or the actual neurological machinery with which the computations are carried out.

For instance, once we have a computational theory of the eye, once we know that the eye's biological function is to discriminate objects in the environment, we can begin to investigate the algorithms and representations needed to perform that function. The computational theory is supposed to feed into level 2 by making available diverse design constraints on the features of cognitive capacities through a systematic task-analysis of the problems the capacity solves. Once we have a detailed enough understanding of the algorithms and representations, we can finally investigate the neurological machinery that actually implements the computational theory.

The notion of functions, therefore, belongs to the common toolbox of theorising in cognitive science. But even though many psychologists agree that *some* understanding of the function of a cognitive capacity is necessary for psychology, not many of them believe that their research project can be based exclusively upon the *evolutionary understanding* of function (that is, a historical understanding concerning the past environments of ancestors). In fact, Marr (1982) himself, without excluding evolutionary function of capacities on principle, thought that investigating current functions of capacities in environments was sufficient for an understanding of how the mind works. The understanding of evolutionary function might prove necessary for a *complete* understanding of the mind as a biological phenomenon, but it should not unduly limit research in psychology.

The majority of cognitive psychologists have followed Marr's reasoning and have, therefore, opted for non-historical accounts of function that look at the performance of cognitive capacities in the *current environment*, rather than the *past, ancestral* environment. One typical argument (exemplified in Lawrence Shapiro and William Epstein (1998)) raised against evolutionary psychology is that psychology cannot be based *exclusively* on this consideration of function derived from evolutionary theory. The underlying reason is seemingly that the concentration on past, ancestral environments will stifle progress in psychology because it will lead psychology unto unsure fields of speculation about how the past might have been.

But, in any case, it is mistaken to see evolutionary psychologists as advocating such a strong view. As John Tooby and Leda Cosmides (1998) have themselves correctly responded, they have always intended the evolutionary perspective to supplement and expand the perspectives already in place in traditional psychology rather than displacing them altogether. It was always clear to them that evolutionary psychology as a psychological discipline does participate in the broader methodological decisions of cognitive science. Evolutionary psychologists believe, though, that in the future this addition of the evolutionary perspective will prove decisive in devising far better experiments that will help scientists to find unknown cognitive devices.

Such misunderstandings of evolutionary psychology can therefore easily be dismissed. Evolutionary psychology, however, faces other problems pertaining to its methodology and its dependence on evolutionary function. One such problem concerns specifically evolutionary psychology's explanatory project.

(B) Objections to the Explanatory Project

As we saw when first introducing the explanatory project in chapter I, the project starts with prior knowledge of a cognitive capacity and then tries to clarify the details of it under the assumption that the cognitive capacity is an adaptation to solve specific problems. The postulation of an adaptation, as we also saw, implies a host of commitments. For to be identified as an adaptation a cognitive capacity has to fulfil criteria such as reliability, efficiency, and complexity (Dawkins, 1986, 1989; Williams, 1966). Obviously, any defence of the explanatory project with its commitment to *complex cognitive adaptations* will have to outline the notion of the required complexity. One attempt at such specification by Steven Pinker and Paul Bloom (1990) emphasises the fact that a complex cognitive capacity in the right sense will be composed of a number of functionally discreet sub-parts that interact with each other in non-random ways to perform the capacity's evolutionary function. The eye provides a clear example of a trait falling under this definition: it is built of functionally discreet sup-parts – some cells detect colour, others detect spatial orientation, etc. – that interact with each other in non-random ways to perform the eye's visual function.

Todd Grantham and Shaun Nichols (1999) take issue with this definition of adaptive complexity in their critique of the explanatory project. To prove that a given cognitive capacity has the right kind of complexity for evolutionary analysis, researchers need to show that it involves interactions among functionally discrete sub-parts. Now, Grantham and Nichols (1999) would agree that a demonstration that the appropriate complexity is realised in a cognitive capacity is sufficient to show that it is indeed a cognitive adaptation in the sense intended in evolutionary psychology, but Grantham and Nichols also believe that current research in cognitive science is not able to single out *enough* such cognitive capacities to make the explanatory project of proving *massive* modularity worthwhile – or even reasonable. What are Grantham and Nichols' reasons?

Recall that the explanatory project starts from capacities that are independently known to exist prior to the application of the evolutionary perspective. To mount an argument in support of the claim that our mind consists of countless cognitive adaptations, evolutionary psychologists will need to wait until cognitive scientists provide them with a much more detailed picture of the functionally discrete cognitive sub-parts that make up the human mind. So far, Grantham and Nichols (1999) point out, cognitive science has only given us at best a clear understanding of complexity in the cases of language and certain perceptual faculties. There are as yet, Grantham and Nichols claim, only a few cognitive capacities that have been shown to have the right kind of complexity evolutionary psychologists are after. But much more evidence is needed to mount a successful argument for *massive* modularity. Current research in cognitive science cannot therefore license any substantial claims about the adaptive cognitive architecture of the human mind (or any other mind, in fact).

Paul Davies (1999) defends a similar but even stronger line. For Davies, the aim of cognitive science is to chart our mental equipment. Since the explanatory project already presupposes knowledge of the relevant cognitive capacities and how they work, evolutionary psychology's explanatory project seems to be redundant from the perspective of cognitive science since cognitive science's aims are already satisfied: the mental equipment is already charted. Davies does not say that results due to the explanatory project are not interesting and cannot contribute to a complete understanding. However, for him these results do not seem to be important enough in furthering the proper aims of cognitive psychology. So it is not easy to see what the explanatory project of evolutionary psychology adds to the aims of cognitive psychology.

The worries underlying both these criticisms are that pursuing the explanatory project is gratuitous and might deflect psychologists' attention away from vital research. But I do not agree with these general statements. The explanatory project and its insistence on evolutionary function is likely to prove a source of valuable insights into the detailed and intricate mechanisms of cognitive capacities. So even though it is true that the explanatory project starts with prior knowledge of cognitive capacities, this knowledge may be of a very general and abstract kind. The explanatory project and its analysis of the evolutionary function of a capacity may then provide further design constraints that can be fed back into the prior knowledge to work out the nitty-gritty detail of the information-processing mechanisms. Some features of those mechanisms, for instance, which were treated as mere noise from the perspective of the prior conception, could turn out to be important features from the new perspective. There seems no reason to deny that the newly attained understanding of design constraints due to the unravelling of the evolutionary function of a capacity furthers our understanding of the capacity itself. Recall the case of pregnancy sickness (Profet, 1992). There, too, the phenomenon of pregnancy sickness was known to exist prior to the onset of an evolutionary analysis. But only the evolutionary analysis of the pregnancy sickness could reveal that the phenomenon is a sensible defence mechanism to protect the developing embryo against various plant toxins. Once the evolutionary reason of pregnancy sickness is known, scientists can start to analyse all the nitty-gritty details of how female bodies detect the deleterious substances.

In an attempt to contest a response such as this, Davies (1999) claims that the insistence on evolutionary function will lead psychologists to systematically miss important insights from studies about subjects afflicted by pathological conditions (such as brain-damaged individuals). The reasoning behind this claim rests with the understanding of evolutionary function. The evolutionary function of a trait is supposed to focus on those behaviours of a trait that are beneficial to the organism, and those beneficial consequences will occur only if the trait behaves normally. Since pathological situations, however, are highly abnormal and thus do not amount to selectively efficacious situations, Davies thinks that accounts based on evolutionary function will tend to be blind to insights garnered from such situations.

But Davies' argument here is simply another instance of the misunderstanding of the explanatory project of evolutionary psychology. It is not the case that evolutionary psychology focuses its attention on the evolutionary function of cognitive capacities to the exclusion of everything else. The account of evolutionary function is intended to augment the available sources to cognitive psychology, not to exclude them. There is thus no reason on principle why evolutionary psychologists could not integrate insights derived from pathological situations to further their understanding of a particular cognitive capacity. Evolutionary psychologists will insist, on the contrary, that the new source of evolutionary function is likely to give us strong insights into the workings of cognitive capacities. Moreover, they might even argue that evolutionary psychology can actually illuminate certain forms of pathology like mental illness (Murphy and Stich, 2000).

Grantham and Nichols' (1999) point that the explanatory project cannot prove that our minds contain large numbers of cognitive *adaptations*, however, is not so easily dismissed. Current cognitive science does not indeed provide us with enough cognitive capacities with the right kind of complexity to support any such proof. But one might say that it is unfair to evolutionary psychology to demand that the explanatory project alone can provide the proof of massive modularity.

On the one side, the explanatory project is not designed to fulfil the overall project of evolutionary psychology all on its own. Rather, the explanatory project is to be supplemented by its reverse: the predictive project that moves from adaptive problems to previously unsuspected cognitive capacities. Evolutionary psychologists believe that once these two projects are put to use, the predictive project will supply more and more cognitive capacities with the right kind of complexity, which the explanatory project can then take on and perform its task on them. Given enough data, the feeding into each other of these two research projects might finally prove something about the adaptive structure of our minds. Until the final data are in, it is hard to see how successful Grantham and Nichols' (1999) point against evolutionary psychology is.

On the other side, the methodology of evolutionary psychology, to which both the explanatory and predictive project belong, is supported not because it provides grounds to accept the whole of evolutionary psychology's massive modularity. As I already noted in chapter I, the methodology of evolutionary psychology is best viewed as a particular way of deriving hypotheses about the mind's cognitive structure, and not itself a statement about that cognitive structure. It is the other theoretical arguments that are meant to support the massive modularity of the mind. The criticism of evolutionary psychology through its methodology is thus unfair and can be dismissed on the grounds that it misunderstands the overall argumentative role the methodological argument plays in evolutionary psychology.

Let us now turn to the predictive project and the criticisms raised against it.

(C) Objections to the Predictive Project

The predictive project, as it was introduced in chapter I, starts from adaptive problems and subsequently predicts the occurrence of cognitive capacities that deal with those problems. A corollary of the project is that scientists must first create a taxonomy of adaptive problems on the basis of which the predictive project can then start. The very possibility of such a taxonomy is a source of intense debate regarding evolutionary psychology.

i) Doubts about building a taxonomy of adaptive problems

Any taxonomy of adaptive problems will have to be founded on knowledge about the past environments and the problems these past environments posed for ancestral species. Davies (1999), Gould (1997a, 1997b, 2000), Stotz and Griffiths (2001), among others, however, argue that one initial problem besetting the predictive project is that knowledge about adaptive problems in past environments is hard to come by in comparison to knowledge about problems in our current environment. These criticisms are motivated by an underlying worry that the predictive project will saddle psychologists with an evidential burden of reconstructing the past which will hamper psychology's overall progress. Gould (1997a, 1997b, 2000) even claims that statements about past environments fall outside the scope of science because such statements cannot be tested in principle, and are thus topics proper to speculation and not to science.

But such negative assessment of our knowledge about the past does not seem reasonable. The reconstruction of past environments on the basis of archaeological, palaeontological, (comparative-) psychological, or genetic records is extensive (Park, 1996).

The analysis of tools made of stone and bone that our ancestors used and the different ways of producing these tools, for example, allow archaeologists to draw inferences about the mind our ancestors most likely possessed that allowed them to produce such tools (Mithen, 1996).

Another example, the microscopic analysis of the tear and wear of the tools may lead to a better understanding of our ancestors' diets: was it based mostly on plants, fruits, or meet?

The study of bones and skeletons of early humans, another example, is a rich source of information on which palaeontologists draw to better understand the structure of humanoid groups (Park, 1996). The slight dimorphism in the anatomy among females and men, when compared with other primate groups, suggests a mildly polygynous mating system.

Moreover, the forensic study of parts of the skeletons can help palaeontologists to chart common ailments, pathogens, and predator attacks our ancestors suffered.

Comparative psychologists, another example, use their knowledge of different primate social groups to highlight the need of early humans living in socially stratified communities to find effective strategies of coalition building and ally formation to protect themselves from the dangers lurking in social environments.

Geneticists, a final example, can tell from genetic samples of populations living in different parts of the world not only when a specific mutation first occurred, but also where it first originated and how it later spread. Geneticists can thus analyse the migration of early human populations on the basis of particular genetic mutations (Luca Cavalli-Sforza, 1997).

This consideration seems to militate against the utterly negative assessments in Gould et al.'s criticism. The list of adaptive problems scientists can put together from archaeology, palaeontology, comparative psychology, and hunter-gatherer studies is wide-ranging.

One further criticism is that conjectures about adaptive problems are not easily amenable to experiments (Stotz, and Griffiths, 2001). We cannot perform controlled and repeatable experiments in past environments that give us an understanding of the specifics of an adaptive problem that influenced the evolution of a particular trait. Stotz and Griffiths (2001) are partly right to highlight this point. But even though it is obviously true that psychologists cannot go back to the past to perform the experiments, psychologists may nevertheless gain insights into the past by investigating current environments – *given that current environments and past environments are sufficiently similar in important ways.* If psychologists were justified in assuming that the similarity between past and present environments holds, they would be able to project insights gained in current environments back to past environments. In any case, it is worthwhile to note that psychology is not the only science that runs into problems with knowledge that cannot be verified directly by experiments. Other perfectly respectable sciences are in a similar position of not being free to do experiments – e.g. astronomy. Evolutionary psychology does not seem to be in a worse position when compared to these sciences. So why do the critics think that evolutionary psychology has a special problem?

The critics maintain that the commitments to past environments are particularly problematic for evolutionary psychology. Ironically, they can even point to passages written by evolutionary psychologists themselves in support of their criticism (Stotz and Griffiths, 2001). Evolutionary psychologists such as Cosmides and Tooby (1994a) are on record as stating that we should not expect a priori that the current environment resembles the past environment in any great detail. Since current environments have changed dramatically during the last 10'000 years or so, partly due to the invention of agriculture, many mismatches are bound to occur between the two environments. The mismatches, in turn, may disrupt the working of many cognitive adaptations in the current environment so that it is not easy to draw any conclusions from the observable behaviour to the actual underlying cognitive capacities. Such systematic scepticism about the relevance of current environments, critics maintain, makes it very hard to establish hypotheses about adaptive problems. For it seems that one of our most important clues to these problems and hence to an understanding of cognitive adaptations is the current environment (Davies, 1999; Grantham and Nichols, 1999).

The best way to clarify the critics' argument is to say that the argument depends on a strong reading of evolutionary psychologists' claim that past environments do not always match current environments in sufficient detail. According to the strong reading, the possibility of a mismatch between past and current environment leads to the conclusion that behaviour in the current environment is ubiquitously maladaptive and cannot therefore be used as a reliable guide to adaptive problems.

However, I think this reading is unwarranted. A better reading is a weak reading of evolutionary psychologists' claim. According to the weak reading, the claim acts merely as a warning to scientists not to treat behaviour that is adaptive in current environments as an *error-proof indication* of underlying cognitive adaptations. More research has to be done to clearly verify that the adaptive behaviour is also relevantly similar to the behaviour that originally led to the development of the underlying adaptation. For it should never be excluded on a priori grounds that mismatches between past and current environments can occur that disrupt the presumed correlation between adaptive behaviour in the current environment and cognitive adaptation. There are at least two reasons to prefer the weak to the strong reading:

1. Unless most features of a living species' past environment correspond closely to features of its current environment, the species would have gone extinct. For if the current environment differs too much from the past environment, the cognitive capacity's (evolutionary) function most probably would have failed to promote the organism's successful reproduction and survival. Given sufficient time, the species would have been displaced by its more efficient competitors. So given the species is still alive, pervasive and widespread differences between past and current environments are unlikely.

2. Most of the things our ancestors did in their past environment, we are still doing in our current environment. We still look out for mates, interact with kin, forge friendly links with strangers, eat particular kinds of food, and communicate with others. Some things

just do not seem to change very much. Based on this kind of consideration, some scientists, for example Charles Crawford (1998), have argued that we should accept a kind of null hypothesis that the relevant past environments and current environments do not differ in important ways from each other. Obviously this point does not imply that changes never take place between past and current environments. But such changes may be harmless as long as they concern only the concrete realisations of an abstract type of information, and not the abstract type of information itself (Crawford, 1998). An example is necessary to clarify what is at issue here. It is reasonable enough to suppose that the actual languages our ancestors used differed in some interesting ways from the languages we use today (in terms of articulation, vocabulary, complexity, etc.). But even though the two kinds of languages differ from each other, they may still realize the same abstract type of linguistic information to which our evolved cognitive capacities are attuned. This is so because the cognitive capacities are not dependent on specific linguistic input, but on input of a specifically linguistic type. So long as the different linguistic inputs belonging to different languages share the same basic structure, the function of the cognitive capacity will not be disrupted. The same kind of reasoning can be extended to other problems organisms have had to face.

The advantage of the weak reading over the strong reading is that it clearly allows the use of data from the current environment to gain knowledge about the past environment – under the condition that scientists verify that no disrupting mismatches between past and current environments have occurred. Given the above consideration supporting this reading, the weak reading seems to be entirely reasonable.

As the above discussion shows, neither the worries about our knowledge of the past nor the possibilities of a mismatch between past and current environments are capable of stalling the project of creating a taxonomy of adaptive problems on which to run the predictive project. First, there are ample resources on which scientists can draw to reconstruct the past, and second, it is unlikely that the differences between past and current environments will be as pervasive as critics seem to suggest. Though both of the criticisms may force evolutionary psychologists to be more explicit in their reasoning, none of the criticisms can call into question the viability of the predictive project so far.

ii) Doubts about deriving cognitive capacities on the basis of the taxonomy

Nevertheless, building a taxonomy of adaptive problems can only be a first step. Based on the taxonomy, evolutionary psychologists intend to derive detailed hypotheses about cognitive capacity. Critics, however, have given reasons to doubt whether this is even possible.

Grantham and Nichols (1999) worry that even the most complete characterisation of adaptive problems will not generate sufficient design constraints to help psychologists derive detailed enough hypotheses about the designs of cognitive capacities. Grantham and Nichols concede that if only we were justified in expecting *optimal solutions* to adaptive problems, this might narrow down the range of possible solutions to a few candidates – namely the ones that approximate the optimal solution to the problem. But they maintain that optimality is a non-starter in evolution.

Their reasoning points to the different evolutionary forces besides natural selection in order to show that optimal solutions are not easily achieved in evolution. There are, for example, diverse genetic, developmental, and cultural constraints that interfere with natural selection and affect it in ways not conducive to the achievement of optimal solutions to adaptive problems.[29] Grantham and Nichols argue that once optimality is undermined, the range of possible cognitive capacities that can solve a problem increases significantly. Moreover, the wider the range of possible cognitive

29 I will discuss these different evolutionary forces in more detail in section 2.of this chapter when discussing a criticism from Gould. For further details on these different forces see the relevant section.

capacities, the more the credentials of the predictive project are thrown into doubt. For with too many possible cognitive capacities opening up, the project cannot make detailed enough predictions about the designs of cognitive capacities. Thus if any predictions are forthcoming under such adverse conditions, these predictions can only be alarmingly vague.

It is true that, theoretically speaking, evolutionary psychologists may run into such difficulties. But there are reasons to think that critics go too far in their worries. Recent developments in behavioural ecology have seen the emergence of more sophisticated optimality models that allow the derivation of quantitative adaptive hypotheses (Krebs and Davies, 1993; Pyke, 1984). Optimality models, generally speaking, try to find an optimal solution to a given problem once all the constraints are taken into consideration. An optimal diet for a bird, for instance, will have to take into consideration the energy expenditure of foraging and searching for prey, the energy value the different preys contain, the frequency with which it encounters its different preys, the possibility of predator attacks on the birds, etc., etc. More importantly, these optimality models are designed to incorporate many constraints – evolutionary as well as developmental – under which a particular species develops its behavioural strategies. Optimality is never an absolute concept – optimality whatever the constraints. Rather, optimality is always relative to a set of given constraints. An optimal solution will thus always incorporate the prevailing constraints under which a given species has to live. There is no reason why evolutionary psychologists could not avail themselves of such sophisticated optimality models to constrain tightly the possible solutions to any adaptive problem.

Additionally, it must also be noted that evolutionary psychology is not confined to working solely with optimality models. It can employ whatever resources it might find suitable to frame its hypotheses – such as hunter-gatherer studies, game theory, paleoanthropology, and primatology. In particular, the comparative data from cross-species comparisons might prove essentially helpful in framing fruitful hypotheses about our cognitive capacities and their evolution (Byrne, 1996; Hauser, 1996; Park, 1996). Comparisons of human capacities with primate capacities, for instance, can provide a rich source of constraints on hypotheses about the features of human cognitive capacities: certain alternative solutions to an adaptive problem may be ruled out at the outset because of our evolutionary heritage.

However, it is true that even this additional consideration will not be able to exclude all but one cognitive solution. Any adaptive problem may allow different alternative solutions that solve the problem equally well. That is precisely the reason why evolutionary psychologists need experiments: only experiments can tell which one of the alternative solutions is realised in a given species. But the evolutionary-based consideration may nevertheless inform those experiments that ultimately decide which cognitive devices the species realises. Again there is no in principle reason why this approach should not work. My suspicion is that, as in the other cases, the intuitive appeal of the criticism derives from an overly strong reading of evolutionary psychology's insistence on adaptive problems. As soon as one starts to catalogue all the resources available to evolutionary psychology, these criticisms lose much of their intuitive appeal.

A final criticism of the predictive project comes from Davies (1999) and Stotz and Griffith (2001). Both of these critics argue that to know anything substantial about the adaptive problems a particular species confronted in its evolution, scientists will have to know something *in advance* about both the array of cognitive capacities organisms used in their interaction with the world and the structure such cognitive capacities instantiated. Without the benefit of such prior knowledge, scientists will not even be able to reconstruct the ecological niche of particular species in the precise way necessary for an evolutionary understanding of the species. This is so because the designs of cognitive capacities are one important clue to the kinds of problems the species had to solve. According to these critics the scientists' ability to model the adaptive problems and their concomitant selection pressures is directly dependent on current descriptive knowledge of the structure of the relevant cognitive capacities. But the predictive project is not able to furnish scientists with such prior knowledge about cognitive capacities because it is the predicative project's aim to first attain this knowledge on the basis of knowledge of adaptive problems.

A concrete example will clarify the criticism. Fish possess what is called a "conspicuous lateral line" (Williams, 1966). This organ is a readily observable morphological trait of most fish that develops by default in individuals, and shows high structural complexity – all of which are properties that indicate the presence of an adaptation. The problem was that scientists investigating the lateral line could not find any plausible evolutionary function for it. Only after detailed investigations were they able to figure out that the lateral line is a sense organ related to audition. They did not immediately hit on the correct solution, because they had to learn that sounds constituted an adaptive problem for fish. It was only on the basis of their prior knowledge about the detailed structure of one morphological trait that scientists were subsequently able to chart the details of this adaptive problem constitutive of the ecological niche of fish. Suitably generalised, the same reasoning suggests that any precise knowledge about the ecological niche of a particular species will have to depend similarly on prior knowledge of the structures of the traits adapted to that niche. That is just the point critics highlight in their reasoning.

Should the criticism turn out to be correct, the predictive project could not even start singling out adaptive problems since it lacks the required prior knowledge about cognitive capacities. I believe this point is well taken. In many cases scientists would not know where to look for adaptive problems unless they had prior knowledge of cognitive capacities guiding them in their search. Moreover, if evolutionary psychology depended solely on the predictive project to derive cognitive capacities, evolutionary psychology would be in serious trouble. But I think that it is a serious misrepresentation of evolutionary psychology to hold that it depends exclusively on the predictive project. The predictive project is only one aspect of evolutionary psychology. Evolutionary psychologists intend the predictive project to be supplemented by the explanatory project. As we saw earlier, the goal of the explanatory project is precisely to give scientists prior knowledge of some cognitive capacities. It may be true that in most cases this prior knowledge will come from outside of an evolutionary consideration proper, i.e. from traditional cognitive science. But it need not always be the case.

Comparative studies based on similarities among evolutionarily related species may prove to be a rich source of prior knowledge of cognitive capacities constitutive of the explanatory project. The reliability of the comparative studies depends on the validity of the hypothesis that evolutionary related species are likely to possess interestingly similar cognitive capacities. Given the validity of the hypothesis, a close consideration of homologous traits shared by closely related species, for instance, can enrich our knowledge of cognitive capacities in one of the compared species without thereby already implying that we have to know all the details of the adaptive problems the species confronts. Our new knowledge derived from comparative records will subsequently feed into the explanatory project. Once a description of a cognitive capacity finds its way into the explanatory project, it will be available to single out the detailed adaptive problems necessary for the predictive project – as the critics suggest.

An additional advantage of the comparative method is that once the comparative data is accepted, the search for adaptive problems is no longer confined to one species alone in isolation of all the others. Given that evolutionarily related species share homologous traits and tend to inhabit similar environments of evolutionary adaptedness, any insights about the adaptive problems of one species may be transferred to the other species as well. The comparative method thus acts simultaneously as a source of knowledge about *cognitive capacities* as well as *adaptive problems.*

So once the comparative method and its results are in place, evolutionary psychologists will be able to successfully address the critics' arguments. Knowledge derived, for instance, from comparative studies are continuously fed into a cycle of complementing research projects in evolutionary psychology that will finally allow a detailed understanding of the adaptive problems a species confronts in its environment. The strategy has wider application. Evolutionary psychologists can break into the mutual dependence of adaptive problems and the cognitive capacities solving them by taking advantage of every source of information available to them. These sources support evolutionary psychologists in either finding independent evidence for the presence of a cognitive capacity *or* the occurrence of an adaptive problem. What the criticisms show at most is that the earlier optimism of evolutionary psychology might have been exaggerated. The relations between the structure of a cognitive capacity and its adaptive problem seem to be more complex than evolutionary psychologists had initially imagined. But they are not successful in calling into question the basic project of evolutionary psychology.

In this section I have shown the criticisms against the methodology of evolutionary psychology to be groundless. Evolutionary psychologists have ample resources at their disposal to respond to the critics' complaints. I next turn to the criticisms aimed at the core concepts of evolutionary psychology.

1.2 Objections to the cognitive architecture and its assumptions

In addition to evolutionary psychology's methodology, the core concepts of evolutionary psychology have increasingly become the target of criticism. Thus some critics have tried to attack the importance of the biological notion of adaptation for the study of the human mind, other critics have suggested that evolutionary psychology has to include other evolutionary forces apart from natural selection, and still others have claimed that the concept of modularity used in evolutionary psychology is misguided. In the following I will draw attention to these criticisms and show how they can be circumvented.

(A) Gould on Adaptation

As we saw in chapter I, evolutionary psychologists believe that the mind is massively modular, that it is a system of cognitive adaptations to solve particular problems. The only known process capable of producing complex adaptations is generally assumed to be the process of natural selection (Dawkins, 1986, 1989; Williams, 1966). Everyone agrees with the last claim. But not everyone agrees with the additional claim that the mind actually contains thousands of cognitive adaptations.

Stephen Gould (1997a, 1997b, 2000) has been one of the most persistent critics of evolutionary psychology and its massive modular mind. He has continuously charged evolutionary psychologists on account of their exclusive reliance on natural selection. Gould has argued in his own work for an alternative perspective that views other forces besides natural selection as equally important. He sees at least three:

The first force is analysed by population geneticists. Recent discoveries in population genetics have shown that much genetic variability is due to neutral, and therefore non-adaptive, changes in the sequence of the nucleotides, which compose a gene (Ayala, 1982). The genetic code allows for particular nucleotides to be substituted with alternative ones without thereby changing the structure and the function of the protein expressed by the genes. Substitution of one nucleotide for another within a natural population may therefore not have adaptive consequences at all, because there will be no selection pressures either for or against it. A mutation may thus spread through the population and perhaps even randomly increase its frequency over generations. Random events like these might alter a population's genetic composition, thereby forcing an evolutionary change upon the population, but still be non-adaptive. Contingency and chance, Gould says, play a crucial role in the history of life, a role that has not been readily acknowledged by evolutionary psychology so far.

The second force affecting evolution is analysed by developmental biologists. Gould points out that researchers have found surprising conservation of basic pathways of development across phyla that have been evolving independently for a very long time. These basic pathways act primarily as important constraints upon the evolution of adaptations and they structure the directions and potentials of future evolutionary change. For evolution has to work on the material it finds available, however imperfect or cumbersome these materials may prove to be for the development of new adaptations. According to Gould, such developmental limitations of new possibilities are a dominant and pervasive theme in evolution.

The third force is analysed by palaeontologists. In competition with the Darwinian model of speciation that views the origin of species as a slow and steady transformation of populations by small genetic changes in individual organisms generation after generation, Gould has proposed an alternative theory of "punctuated equilibrium". The theory of punctuated equilibrium rests on two observations drawn from the palaeontological record. First, most new species originate in a geological "moment". The fossil records show a relatively sudden appearance of newly developed species. Second, species generally do not change in any substantial or directional way during their geological reign. According to punctuated equilibrium, the history of life is characterised by long and relatively stable evolutionary periods that are interspersed with sudden occurrences of mass extinctions or other dramatic events. It is such punctuating events and not the small incremental genetic changes that drive the speciation of populations and that are responsible for the sudden occurrence of new species often observed in the palaeontological records. According to Gould, therefore, the Darwinian model of speciation is misleading, and as long as evolutionary psychology continues to pay allegiance to it, evolutionary psychology too will be misleading.

Based on his worries about the exclusive role natural selection plays in current evolutionary theorising, Gould concludes that the notion of a massively modular architecture that contains literally thousands of cognitive adaptations must be deeply misguided. He grants that our "big brains" will likely turn out to be an adaptation to the complexities of hominid social living, but he flatly denies that our brains contain specialised adaptations to solve the particular problems our ancestors faced in their environment. Consequently, many features of human cognition will, Gould thinks, probably turn out to be non-adaptive products of a general intelligence that evolved for different adaptive reasons.

I must pause the argument here because to understand Gould's points correctly, one must first distinguish clearly between three different origins of cognitive capacities. I have already touched upon these distinctions in the last chapter, but I need to recall them here to elucidate Gould's points. An investigation of the origins of cognitive capacities allows us to draw the following distinctions:

Adaptations
Traits that originated by natural selection for a particular purpose and which still perform that same purpose in the current environment (Dawkins, 1989; Gould, 1982; Williams, 1966).

Exaptations
Traits that originated for some particular purpose, but which were later co-opted to perform a different purpose (Gould, 1982).

Spandrels
Traits that originated as mere by-products of architectural or developmental constraints and which had no adaptive purpose at first, but which might be later co-opted to perform such an adaptive purpose (Gould, 1979).

Gould believes that our mind does not contain cognitive *adaptations* that originated by natural selection (apart from the big brain), nor does he believe that the mind contains many cognitive capacities that originated to perform a particular adaptive function, but which are now performing different functions altogether (exaptations). Instead, Gould believes that our mind most probably contains lots of spandrels that originated as non-adaptive by-products due to the architectural constraints of building a big brain like ours. Most of these spandrels have later been co-opted to perform certain functions in our evolutionary history, of course, but strictly speaking they are not adaptations or exaptations, even if the co-option has been effected by natural selection after they originated. Here is Gould in his own words on the topic:

> If any organ is, prima facie, replete with spandrels, the human brain must be our finest candidate -- thus making adaptationism a particularly dubious approach to human behavior. I can adopt (indeed I do) the most conventional Darwinian argument for why the human brain evolved to large size – and the nonadaptationist principle of spandrels may still dominate human nature. I am content to believe that the human brain became large by natural selection, and for adaptive reasons -- that is, for some set of activities that our savanna ancestors could only perform with bigger brains. Does this argument imply that all genetically and biologically based attributes of our universal human nature must therefore be adaptations? Of course not. Many, if not most, universal behaviors are probably spandrels, often co-opted later in human history for important secondary functions. The human brain is the most complicated device for reasoning and calculating, and for expressing emotion, ever evolved on earth. Natural selection made the human brain big, but most of our mental properties and potentials may be spandrels -- that is, nonadaptive side consequences of building a device with such structural complexity. (Gould, 1997b, p. 37)

Keeping the fine distinctions between the origins of cognitive capacities separate, one can appreciate the radical thesis about the structure of the mind that Gould suggests. In accordance with his views, Gould has persistently accused evolutionary psychology of over-extending adaptations in a Panglossian manner: whichever trait scientists investigate, adaptations are abounding. He has further accused evolutionary psychology of systematically disregarding the pervasiveness of contingency in evolution. Tooby and Cosmides (1997) have enumerated the different charges Gould has been raising in his writings against them thus:

> In particular, any reader of English would understand Gould to be specifically claiming that we are panglossian panadaptationists (although perhaps retrospective panglossians), are either unaware of or do not use neutral theories of evolution in our work, never consider or test byproduct hypotheses, are unaware that byproducts and constraints exist and are ubiquitous in the design of organisms, are befuddled by the pervasiveness of contingency in evolution, spend our time concocting 'untestable' and hence 'unscientific' post hoc just so stories, and so on. (Tooby and Cosmides, 1997, p. 3)

How justified is Gould in his assumption that evolutionary psychologists are relying exclusively on adaptations and natural selection in their theorising? Tooby and Cosmides (1997) believe that he has completely misconstrued their approach to human cognition. In their reply to Gould, Tooby and Cosmides (1997) emphasise that the exclusion of rival hypotheses to the adaptive hypothesis is a matter-of-course in evolutionary biology and therefore should be so treated in evolutionary psychology too. In order to show that Gould's arguments are groundless they cite Leda Cosmides' (1989) own work on the Wason selection task as an example. The Wason selection task tests people on their performance on conditional rules of the form: if p then q.[30] In her own work, Cosmides (1989) started

30 The Wason selection task is a classical test of conditional reasoning. In the test, subjects are given a conditional rule, if P then Q, and are asked to select which of four given cards must be turned over to find out whether the conditional rule holds or not. Each card has the information regarding whether P holds on one side, and whether Q holds on the other side. A typical subject is thus shown four cards. For example:
Conditional: If P, then Q
Cards: [1] P [2] Not-P [3] Q [4] Not-Q
Each card states on the back whether the other condition holds true or not. The correct response in accordance with the rules of logic is to turn over the two cards:

from the well-known fact that subjects are much better in reasoning about conditionals when social rules with concrete contents are involved.[31] Cosmides (1989) proposed an explanation of why subjects improve their reasoning when social rules are involved. Her explanation rests on the hypothesis that our minds are equipped with cognitive adaptations with content-dependent principles created by natural selection that are specifically attentive to the social features of relevant situations. Once social situations of the relevant type are detected, the respective cognitive adaptations take over the reasoning and lead to the improved performances on the Wason selection task. Because we are not as well supposed to be equipped with cognitive adaptations for abstract reasoning tasks stated in the forms of simple if p then q, the negative performance on the Wason selection task are to be expected since we cannot benefit from content-dependent principles due to cognitive adaptations.

[1] P [4] Not-Q

They are the only cards that can definitely prove the conditional wrong depending on what information is printed on the other side. Results show that when the test is put in purely abstract terms of P and Q, subjects perform rather poorly (typically they do not select the Not-Q card, and most likely select the Q card).

31 It was already known in the literature prior to Cosmides' work that subjects improve considerably their reasoning when conditional rules are re-phrased to contain social rules like the following:
Conditional: If a person has a beer, then the person must be over 18
Cards: [1] Person has a beer
[2] Person hasn't a beer
[3] Person is over 18
[4] Person is not over 18
Then subjects' responses are significantly better: they turn over both "Person has a beer" [1] and "Person is not over 18" [4] more often (which is, of course, the logically correct answer). Since these differences in subjects' performance seem to depend solely on the content of the conditional rule, and not on its logical form – since the conditionals still have the same form if p, then q --, these differences must be due to a *content effect.* Ever since the detection of these content effects, psychologists have wanted to explain how the non-formal features of a problem influence the different test performances observed.

Tooby and Cosmides (1997) claim against Gould that the adaptationist explanation proposed in (Cosmides, 1989) was actually informed by the rejection of alternative hypotheses that apparently fit the bill of non-adaptive cognitive capacities Gould had in mind all along. One such rejected alternative theory, *availability theory*, tries to explain content effects on the grounds of general-purpose cognitive mechanisms by suggesting that subjects' performance on the test depends *solely* on their familiarity with the situation. So if a subject has had past experiences with the content of the conditional rule, her performance is boosted on the test. According to the availability theory, there is consequently no need to postulate specialised modules dedicated to solving particular types of situations. Moreover, while it may be true that general-purpose mechanisms originally developed for some adaptive reasons, their use in particular domains such as social contract situations is probably just a side-effect of these mechanisms being co-opted to perform a given task. Cosmides (1989), however, found that performance on the Wason selection task is not correlated with past experiences of the content domain probed in the Wason selection task.

In any case, this example shows, against Gould, that evolutionary psychologists are well aware of the possibility of rival theories. Gould is correct in chiding evolutionary approaches to the mind that confound the different origins of cognitive capacities. Any theory must be very careful not to over-extend the notion of adaptation gratuitously. But whereas Gould claims that evolutionary psychologists do not test for alternative theses beside natural selection, evolutionary psychologists claim instead that they do. I have presented one example, the Wason selection task, which supports evolutionary psychologists' claim. One should therefore rule out Gould's claims as unfair to evolutionary psychology. But Gould's consideration retains its importance as a warning against an overconfidence in adaptation. Any application of adaptation to the mind or body within psychology must be submitted to experimentally, and it is only experiments that can settle the issue. However, I do not see any principled reasons in Gould's arguments that amount to anything like a conclusive argument against evolutionary psychology.

(B) Miller on Sexual Selection

Like Gould, Geoffrey Miller (2000a, 2000b) argues that evolutionary psychology needs to change its exclusive focus on classical adaptationism. In contrast to Gould, though, who argues for the importance of non-adaptive processes in evolution, Miller wants to broaden the concept of adaptation, which is traditionally based on natural selection alone, to include processes of sexual selection as well. While natural selection depends on the success of both sexes in relation to survival, sexual selection depends on the success of certain individuals over others of the same sex, in relation to propagation. The mechanism of sexual selection is of two kinds: in the one it is a struggle between individuals of the same sex, generally males, to out-compete or kill their rivals in the mating game; in the other it is again a struggle between individuals of the same sex, generally females, to attract those of the opposite sex and to choose the more genetically fit ones. Miller contends that evolutionary psychologists have so far unduly concentrated on cognitive adaptations shaped by natural selection for survival. But he believes, however, that such a concept of adaptation is too narrow to capture other types of adaptations shaped by sexual selection.

According to Miller (2000a), we find a first indication that the classical concept of adaptation is inappropriate to capture the whole range of adaptations in the fact that sexually shaped adaptations fly in the face of many of the well-known criteria for adaptations such as efficiency, complexity, modularity, and universality. If evolutionary psychologists do not want to run the risk of leaving out genuine candidates of adaptations, they will have to broaden their concept of adaptation to include these different adaptations. Miller's argument depends on the exact details of why sexually selected traits thwart the traditional criteria of adaptations. It is therefore necessary to review Miller's points in some detail in order to clarify his conclusions. A good way of reviewing his points is by analysing the kinds of problems the mating game poses in more detail.

Sexual selection originates with mate choice based on reliable signs of heritable fitness (Zahavi, Zahavi and Balaban, 1997). Some features of a trait, for instance, will be favoured in the mating game because those features advertise something reliable about the carrier's "good genes", which he can potentially pass on to its offspring. Sexual selection through mate choice will therefore install a process in which some features of a trait are selected in the opposite sex because they are better adapted to win the mating game. For the choosing party in the mating game, the crucial problem is to find reliable ways of assessing the truthfulness of the fitness indicators the opposite sex parades. Its problem is to choose those organisms out of a group of competitors whose fitness indicators will actually fulfil the promise of heritable "good genes". Unless the choosing party possesses reliable means of assessing the reliability of fitness indicators, it will always pay the carrier of those indicators to cheat and to give only the appearance of being a good mate.

As recent studies in animal signalling theory have shown, sexual selection consistently selects for costly traits that drain the energy resources of the carrier and affect its survival chances in order to reject cheaters with unreliable fitness indicators. High costs incurred by the traits are an effective means of ensuring that the traits are reliable fitness indicators because individual organisms without "good genes", i.e. cheaters, do not have the resources to sustain such costly traits. It is, therefore, precisely the high fitness costs that are needed to sustain these traits that make them sexually attractive for the choosing party because they are reliable fitness indicators.

The peacock's tail, for instance, is a trait that fits all the requirements of a sexually selected trait. The complexity of the structure of the peacock's tail, the brilliancy of its colours, its excessive length in comparison to the rest of the body, all concur to burden the peacock with large fitness costs. The purity of the colours of a peacock's tail is affected by pathogens so that bright colours necessarily indicate a good resistance against attacks by pathogens. The excessive length of the peacock's tail makes its carrier more conspicuous to the eyes of predators so that peacocks with longer tails also run a higher risk of being eaten by predators. But it is precisely these sorts of high fitness costs that make the peacock's tail such a sexually attractive trait for the peacock hen. The more extravagant and elaborate the tails are, the more sexually attractive the carrier becomes for the opposite sex. It is the most extravagant carriers that will enjoy a more than usual rate of reproduction.

The high fitness costs that a sexually selected trait imposes on the carrier, however, look very inefficient from the standpoint of natural selection for survival. Such a standpoint is inclined to focus exclusively on efficiency as one of the criteria for singling out proper adaptations. Thus if the main purpose for an organism were to survive successfully, the costly traits selected by sexual selection would be an oddity because they would be inefficient to guarantee the survival of the individual organism. It would further follow that a standpoint fixed on the efficiency of survival would necessarily miss adaptations due to sexual selection. In the end, efficiency for survival cannot play the same role in detecting sexually selected traits as it does in the context of natural selection.

Similar arguments apply to the other criteria traditionally proposed for singling out adaptations in individual organisms. Complexity, too, can be shown to create difficulties when applied to sexually selected traits of individual organisms. The reasoning behind these difficulties is easy to see once a concrete example of a mate-choice situation is outlined. In many mate-choice contexts, sexually selected traits in one sex promote a carrier's genetic fitness by influencing the cognitive mate-choice mechanisms of the opposite sex. Male frogs, for instance, call out for females by emitting croaks of particular wavelengths that affect the auditory systems of the females. In the given example, we have two complementing mechanisms. First the sender's behavioural adaptations responsible for emitting signals of a particular wavelength. Second the perceptual capacities of the receiver that are involved in the processing and decoding of the emitted signals. If complexity can be applied at all in these circumstances, it will be seen as reasonably applying to the whole signalling system containing *both* the sender *and* the receiver. Complexity need not always be confined to the individual organism. Sometimes complexity must be treated as distributed over more than one individual organism.

Once this fact is appreciated, one can understand the possibility of how the most important aspects of complexity are realised in the perceptual-cognitive mechanisms of the receiver. That is, it could turn out that the sender's behavioural adaptations responsible for emitting the relevant signal are rather simple when compared to the receiver's adaptations. For instance, the male frog's behavioural adaptations that are responsible for producing a croak of a particular wavelength may be much simpler than the female's perceptual mechanisms processing and decoding the signal. From the perspective based on traditional notions of complexity and its focus on individual organisms, these simple but nonetheless sexually selected traits in the sender cannot count as proper adaptations. Their simple nature disqualifies them from the start. The same is not true if scientists analyse the larger signalling system between sender and receiver. Once the whole system comes into view, the required complexity emerges and scientists can treat the whole signalling system as a well designed system for effective signal transmission. Moreover, the new perspective on complexity based on sexual selection permits scientists to discover adaptations even in simple traits if these simple traits contribute their results to a larger system designed for a specific purpose.

On the basis of this consideration pertaining to the other classical criteria of adaptation, Miller emphasises the need for evolutionary psychology to broaden its account of adaptation. Not only must it include adaptations that are the products of natural selection to win the survival game, but it must also include adaptations that are the products of sexual selection to win the mating game. Without these necessary changes, evolutionary psychologists are liable to overlook important adaptations of our human cognitive architecture. It is correct to observe that evolutionary psychologists are aware of these issues, at least in their writing. But they seldom pay more than lip service to sexual selection in their theories. Miller is therefore right to admonish evolutionary psychologists for neglecting sexual selection. Unless evolutionary psychologists include theories informed by sexual selection, evolutionary psychologists will unnecessarily limit their own research. That said Miller's is a sympathetic critique of evolutionary psychology. Its aim is not to uncover a principled failure of the research program, but rather to help it out by showing ways to improve the research program. The next criticism, however, is a more direct challenge to evolutionary psychology.

(C) Karmiloff-Smith on Modules

Most scientists would agree that evolution has given us more and more complex behavioural capacities to better adapt to our demanding environments. But some scientists deny the assumption that cognitive modules in the guise of genetically pre-specified structures influence these complex behavioural capacities. Annette Karmiloff-Smith (1992) claims that such modules are not important in the development of the human mind from infancy to adulthood. According to Karmiloff-Smith (1992, 2000), infants do not first possess pre-specified blueprints of cognitive modules which are then assembled by the developmental process as the right kind of input triggers them online. She postulates that instead of these blueprints, infants possess only a small set of basic and very general cognitive biases in addition to whatever general-purpose learning mechanisms are necessary to promote development. The multitude of discrete domain-specific cognitive capacities later observed in adults grows out of these initial cognitive biases through gradual differentiation and other developmental processes. It is a product of development, and not of genes. Karmiloff-Smith reasons that most of the cognitive capacities that evolutionary psychologists investigate will turn out to be products of ontogenetic development, and not adaptations at all.

Karmiloff-Smith (2000) tries to defend her claim by revaluating some of the most important evidence evolutionary psychologists use to support their own theoretical views in favour of modularity (Botterill and Carruthers, 1999; Hirschfeld and Gelman, 1994). Her strategy is to propose alternative interpretations of the same evidence, but which support her own theory of ontogenetic products. I shall quickly outline the typical evidence evolutionary psychologists use to support their views on the modular mind as these are relevant to understand Karmiloff-Smith's criticism. Most of the evidence is already familiar from our discussion of chapter I. The sources of evidence are threefold:

1. Dissociations:

People with dissociations manifest disrupted cognitive capacities in particular domains, while simultaneously showing cognitively normal performances in other domains. The disruption of cognitive capacities may be due either to genetically induced causes or to brain injuries acquired in adulthood. For instance, subjects with Williams' syndrome, a genetically induced affliction, manifest good performance in face processing, language, and social skills, but show flawed performance in spatial cognition, numbers, and problem-solving skills (Karmiloff-Smith, et al., 1995). In addition to dissociations emerging across cognitive areas, there are also dissociations occurring within one cognitive area. There are, for example, people with visual systems that manifest normal cognitive performances in the area of object recognition, but whose visual systems are at the same time impaired with respect to face recognition.

2. Atypical development during infancy:

People with atypical development during infancy are affected by abnormal conditions during development that selectively impair one cognitive domain, while the rest of the brain is believed to develop normally. For instance, autistic children are believed to suffer from a biological impairment that specifically affects the development of their ability to understand mental states of others (Baron-Cohen, 1995). The autistics' reasoning about physical objects, however, shows normal performance. It is just the abilities concerned with the understanding of mental states in others that are knocked out in development.

3. Precocious abilities in infants:

Recent developmental studies show that infants typically manifest high proficiency in particular cognitive areas from an early age on. At just a couple of months old, for example, newborn babies demonstrate preferential interest in face-like shapes that are highly indicative of domain-specific and modular cognitive capacities (Johnson and Morton, 1991). While the precocious abilities show high level of performances very similar to adult abilities, most other areas have to wait for further experience to attain similar levels of performance.

So while evolutionary psychologists use these three sources to support their views on modularity, Karmiloff-Smith (2000) suggests, on the contrary, that the evidence is likewise open to alternative interpretations that do not allow derivation of modularity. Her alternative interpretations take one of the following forms:

First, the evidence from dissociation turns out on further inspection to be too weak to favour the modularity view over Karmiloff-Smith's view that the modular mind of the adult brain is simply the result of gradual development. To support her claim, Karmiloff-Smith points out that most of the dissociation data stem from adult humans that received brain injuries and not from infants. But it is clearly the data from infants that would be most relevant in showing whether or not we possess pre-specified cognitive modules. The dissociation data from adult humans by themselves are too ambiguous to favour one of the two competing views of the mind. For if

Karmiloff-Smith's view of the mind is correct, dissociations in adults may have affected a mind at a stage where development by general-purpose mechanisms had already installed the different domain-specific capacities. According to Karmiloff-Smith, there is no doubt that the human brain shows a high degree of modularisation in adulthood, but this modularisation is a product of development and not something that is genetically pre-specified in the mind in an initial stage.

Second, evolutionary psychologists have used data from atypical development during infancy to support their views because these data seemingly show that particular cognitive capacities can be selectively impaired while leaving other parts of the brain functioning normally. Karmiloff-Smith (2000) contests this interpretation of the data by claiming that the underlying assumption is wrong that the unaffected parts of an atypically developing brain are functioning normally. According to Karmiloff-Smith, a more accurate perspective of the brain will point to the fact that brains are intricate systems, which are subject to exceedingly complex dynamic processes of development. From this perspective, a brain affected by atypical development is already a brain that develops very differently from a normal brain. One should therefore expect that a brain affected by atypical development will show abnormalities throughout the range of all its cognitive abilities. The question is whether data exist that can demonstrate if unaffected parts of an atypical brain perform normally or abnormally.

Karmiloff-Smith cites her own work on people with Williams' syndrome to disprove evolutionary psychologists' interpretation of the data from atypical development during infancy. As I already outlined above, people affected by Williams' syndrome manifest normal performances in face recognition and language even though other parts of their brain are severely impaired. If face recognition and language really are cognitive modules that work in isolation from other parts of the brain, psychologists should expect that the associated cognitive mechanisms are the same in people with Williams' syndrome and people without it. If, on the other hand, Karmiloff-Smith is right, one should expect to detect significant differences among the cognitive mechanisms used by people with Williams' syndrome and people without it. Supported by the use of brain-imaging techniques and other methods, Karmiloff-Smith (2000) was able to report specific differences in the cognitive mechanisms used by people with Williams' syndrome and people without it. Her evidence therefore suggests that particular disruptions during development are not selective in the ways assumed by evolutionary psychologists. Atypical development of the brain seems to affect cognition in much more complex ways than is often assumed. In view of these new findings, it seems clear that the evidence from atypical development cannot be used to support strong modularity theses in evolutionary psychology.

Third and finally, Karmiloff-Smith contests the interpretation evolutionary psychologists give to the evidence from the precocious abilities of infants. For evolutionary psychologists maintain that the precocious abilities that children manifest are conclusive evidence for cognitive modules. Karmiloff-Smith again cites her own research on the performances of children that seems to contradict the interpretation. Her brain-imaging data demonstrate that specialised cognitive capacities in infants show a gradual development over their development. She shows, for instance, that 6 month old babies initially process face-like inputs using several distributed areas of the brain, and not one particular area dedicated to face-processing. By 12 months, however, the babies' face processing mechanisms are specialised for one domain and much more localised. These results seem to support Karmiloff-Smith's model of the mind over evolutionary psychology's model.

To summarise: Karmiloff-Smith has tried to mount arguments against evolutionary psychologists' appropriation of the three main sources of evidence from development. Her strategy throughout has been to delineate alternative interpretations of the evidence that support her own view, which claims that the modularised brain in adults is a product of development, and not the outcome of genetically pre-specified cognitive modules. Her observations open up interesting empirical topics of future research, and they might force evolutionary psychologists to tighten some of their arguments to defend them against Karmiloff-Smith's attacks. Her observations, however, seem not call into question evolutionary psychology's programme. I believe therefore that evolutionary psychologists have ample resources to invent counter-strategies that defuse much of Karmiloff-Smith's arguments. In the following I shall outline three such counter-strategies evolutionary psychologists may take up against Karmiloff-Smith.

First, evolutionary psychologists can easily undermine Karmiloff-Smith's observations that evidence from dissociation stemming from adult human brains cannot favour evolutionary psychology's modularity views. Their most obvious strategy would be to ensure that their evidence also contains data from infants, preferably very young ones. This alone should be sufficient to reject Karmiloff-Smith's observations – if the data turns out right.

Second, evolutionary psychologists should also reject Karmiloff-Smith's arguments against evolutionary psychologists' reading of the evidence from atypical development. The most effective way for evolutionary psychologists to proceed is to highlight a questionable underlying assumption in Karmiloff-Smith's reasoning. Karmiloff-Smith apparently believes that each cognitive module is (neurologically) localised in a specific brain area. However, nothing forces evolutionary psychologists to commit to such an assumption. Evolutionary psychologists can claim instead that most cognitive modules are distributed across several areas of the brain. Should this turn out to be the case, finding irregular brain activities between typical and atypical brains in one specific brain area could no longer imply that no cognitive modules are present – as Karmiloff-Smith appears to suppose. For given that cognitive modules have a distributed realisation in the brain, some disruptions in one brain area may result in slightly different brain activities within or across different brain areas – some brain areas might try to compensate for the dysfunctional ones. The observed brain activities of the tested specific brain area, however, could show at most that the cognitive module is dysfunctional, but not that there is no module at all. Consequently, Karmiloff-Smith has not given conclusive evidence that there are no modules in the sense explored by evolutionary psychology.

Third and finally, evolutionary psychologists should further reject Karmiloff-Smith's readings of the evidence from precocious abilities in infants. Evolutionary psychologists can choose from two strategies against Karmiloff-Smith's observations.

The first strategy notes that certain modularised abilities are only activated during special pre-specified periods in development. It is now believed, for instance, that children's genetically pre-specified abilities to understand others' mental states are fully expressed only around 3 1/2 or 4 years of age (Leslie, 1987, 1994, 1995; Baron-Cohen, 1995). Some modules, therefore, have to wait for their full expression until the right developmental conditions pre-specified in their genetic programmes are realized. So given the differences that Karmiloff-Smith has found in the face-processing mechanisms among 6 month and 12 month old infants, evolutionary psychologists may try to reinterpret her data accordingly. Instead of saying that the face-processing mechanisms undergo specialisation during development, evolutionary psychologists could say that it is only around 12 months of age that a genuine face-processing mechanism (a face processing module) goes online. Before 12 months of age, face-processing may be underwritten by other mechanisms, probably even general learning abilities that infants have at their disposal. But with 12 months of age, it is the face-processing module that takes over full command.

The second strategy notes that development is possible even in genetically pre-specified modules. In chapter I, section 2., when discussing cognitive modules, I cited Scholl and Leslie as rightly emphasising that "there is nothing in the notion of modularity which prevents even 'matured' modules from developing" (1999, p. 136). Accordingly, Karmiloff-Smith's underlying assumption that modules are completely rigid structures that exclude development is unjustified. So the differences that show up in her data between 6 months and 12 months old babies could be due purely to development *within* a single face-processing module. Again, Karmiloff-Smith's suggestion that babies do not use modules, but instead general learning mechanisms informed by development is not forthcoming.

As it turns out, evolutionary psychologists have sufficient means to reject all of Karmiloff-Smith's arguments, which seemed initially to deliver strong blows against their research project. In the end, evolutionary psychology survives Karmiloff-Smith's attempts at disproving it unscathed.

To sum up this first section, all of the presented arguments so far have touched upon empirical issues concerned with the research project of evolutionary psychology. Some arguments attacked the specifics of evolutionary psychology's methodology by claiming that it is empirically unjustified, the rest of them challenged the concepts of adaptation, natural selection, and modules, which evolutionary psychologists use to support their own views of the mind. While some of the arguments point to weak spots in evolutionary psychologists' reasoning, none of them do anything to call into question the basic tenets of their research project. However, none of the presented arguments have directly engaged with evolutionary psychology's theoretical arguments as first introduced in the last chapter. In the following section I shall present an argument used by Richard Samuels, which is designed to directly engage with evolutionary psychology's theoretical arguments.

2. Samuels' Library Model of Cognition

In chapter I, section 2. I presented evolutionary psychologists' theoretical arguments: the argument from methodology, the argument from the trend to specialisation, the argument from statistical information, and the argument from combinatorial explosion. All of the arguments are intended to complement each other in building a strong case for the claim that once evolution is considered seriously in psychology, the massive modular mind is the most plausible candidate to emerge from it.

Recently, Richard Samuels (1998, 2000) has argued that the theoretical arguments evolutionary psychologists use in support of massive modularity are defective. As will be shown shortly, Samuels' arguments depend on the details of the interpretation of evolutionary psychologists' claims. This is the place where the earlier distinction between a weak and a strong version of evolutionary psychology will pay off. But before I can relate Samuels' conclusions to the different versions of evolutionary psychology in order to evaluate how effective Samuels' arguments are, I will first need to outline Samuels' arguments in detail. I turn to this task now.

According to Samuels (1998, 2000), the most exciting new proposal that evolutionary psychologists are championing is the proposal that the mind is largely or even entirely composed of a distinctive type of computational modules. In order to distinguish these distinctive computational mechanisms from ordinary computational modules, Samuels introduces his term of *Darwinian modules. Darwinian modules* are computational modules that are adaptations produced by natural selection for a specific purpose. According to Samuels' view, then, evolutionary psychologists propose that the mind is largely or even entirely composed of *Darwinian modules*. Additionally, he asserts that evolutionary psychologists reject general-purpose computational mechanisms on account of their evolutionary implausibility. According to Samuels' reading of evolutionary psychologists, they maintain that the logic of the evolutionary process excludes general-purpose computational mechanisms from playing any important part in the cognitive lives of organisms.

Samuels' main strategy in showing that evolutionary psychology's theoretical arguments are not sufficient to rule out general-purpose computational mechanisms as implausible is firstly to introduce an alternative and polemical model of the mind – his library model of cognition (Samuels, 1998) –, and secondly to show that the theoretical arguments of evolutionary psychology cannot favour computational modules over general-purpose computational mechanisms. I will therefore have to initially outline Samuels' library model of cognition, before I can then move on to a characterisation of his arguments proper.

In order to serve its polemical purposes against evolutionary psychology, the library model equips cognition with a general-purpose computational mechanism that works in the context of a complete absence of special-purpose computational mechanisms. However, the library model of cognition does not deprive cognition of all its specialised features. For even though cognition is debarred from accessing computational modules and their specialised procedures, it is nevertheless given access to multiple bodies of domain-specific knowledge that help the general-purpose computational mechanism to solve its imminent tasks. So all specialised content that cognition has at its disposal derives exclusively from the knowledge modules that are installed in the mind by natural selection and other learning procedures. Now that the basic elements of the library model of cognition are outlined, I can move on to the characterisation of Samuels' position which sees the theoretical arguments of evolutionary psychology as not sufficient to favour evolutionary psychology over his library model of cognition. Because the arguments from evolutionary psychology have already been introduced in detail in the last chapter, I shall present them here only briefly.

As we saw in the last chapter, evolutionary psychologists believe on the basis of their argument from the trend to specialisation that a properly understood analysis of adaptive problems leads directly to the postulation of specialised cognitive structures. For given that two adaptive problems have different solutions, it will likely pay the organism to possess two specialised solutions rather than one. From an optimal point of view, then, special-purpose solutions will always be superior to general-purpose ones because the former will outperform the latter in virtue of speed, reliability, and efficiency. According to Cosmides and Tooby, one can easily move from the observation that:

> (1) As a rule, when two adaptive problems have solutions that are incompatible or simply different, a single solution will be inferior to two specialised solutions. (Cosmides, and Tooby, 1994b, p. 89)

to the conclusion that:

> (2) [...] domain-specific cognitive mechanisms...can be expected to systematically outperform...more general mechanisms. (Cosmides, and Tooby, 1994b, p. 89)

Samuels (1998) maintains that this transition is not warranted. He notes that (1) does not specify what cognitive structures would count as "specialised solutions" to adaptive problems: is it bodies of domain-specific knowledge or rather the special-purpose computational mechanisms manipulating them? (2), on the other hand, settles for special-purpose cognitive *mechanisms*.

There are two possible readings of (1). One reading of (1) states that any specialised solution, whatever its nature, will prove superior to general-purpose ones. So although (1) shows, if true, that evolution does indeed favour specialised cognitive structures over general-purpose ones, it does not show that evolution must give rise to special-purpose cognitive mechanisms as opposed to bodies of domain-specific knowledge. Nothing in (1) precludes the possibility of a mind that uses general-purpose mechanisms in combination with bodies of domain-specific knowledge. The insistence on special-purpose cognitive mechanisms in (2) therefore cannot be derived on this reading of (1) alone. Consequently, the library model of cognition cannot be ruled out by (1).

However, another reading of (1) is that the mentioned special cognitive solutions in (1) should be understood as including only special-purpose computational mechanisms. Thus (1) claims that whenever two adaptive problems require different solutions, a single general-purpose *computational mechanism* will be inferior to two special-purpose *computational mechanisms*. According to this reading, (1) and (2) together apparently support the conclusion that domain-specific computational mechanisms will always systematically outperform general-purpose ones. But even then, as Samuels correctly notes, it is far from clear that the argument can support massive modularity over the library model of cognition. For although it might very well be the case that *some* general-purpose computational mechanisms are inferior to special-purpose ones, *other* general-purpose computational mechanisms might perform equally well. It all depends on the kinds of general-purpose computational mechanisms one employs. It is extremely useful that Samuels (1998) distinguishes two very different general-purpose computational mechanisms:

(A) General-purpose computational mechanisms that are not special-purpose, i.e. ones that are not dedicated to solving problems in a single specific domain.

(B) General-purpose computational mechanisms that are not special-purpose – and do not benefit from any innate, domain-specific knowledge or innate, special-purpose mechanisms.

If one means to contrast general-purpose computational mechanisms of type (B) with special-purpose computational mechanisms, it should be clear that they will lose when pitted against the performance of special-purpose computational mechanisms. Since such general-purpose computational mechanisms make no specific provisions for the different kinds of adaptive problems, they will be outperformed by specialised computational mechanisms. But the same does not seem to hold for general-purpose computational mechanisms of type (A). Although these computational mechanisms are general-purpose by nature, they need not treat all problems in the same fashion. They may treat different problems in different ways by accessing different bodies of domain-specific knowledge devoted to specific adaptive problems. Once this is granted, and nothing so far precludes it, one sees that the library model of cognition might prove as powerful as an architecture that contains only special-purpose computational mechanisms. At the very least, it is far from clear that the library model of cognition and its (A)-type general-purpose computational mechanisms would be easily outperformed by special-purpose computational mechanisms. Once the details of Samuels' library model of cognition are known, it is easy to see how the rest of evolutionary psychologists' arguments similarly tumble. I will evaluate these other arguments from evolutionary psychology to show how Samuels deals with them one by one.

As we saw in the last chapter, the argument from statistical information attempts to show that general-purpose computational mechanisms are ill fitted to pick up information that was not observable by an individual during its lifetime because the crucial information would emerge only over evolutionary time (Cosmides and Tooby, 1994b). As Samuels (1998, 2000) remarks, however, even if it is true that general-purpose computational mechanisms on their own cannot learn about such statistical information (=type B), general-purpose computational mechanisms can provide for innate knowledge about the required information by accessing relevant bodies of domain-specific knowledge (=type A) that natural selection has installed in the cognitive architecture. The argument, then, cannot give us a conclusive reason to believe that organisms have special-purpose computational mechanisms that exploit such information rather than general-purpose ones that access specialised knowledge and, hence, we cannot rule out the library model of cognition.

The argument from combinatorial explosion, also introduced in the last chapter, tries to demonstrate that combinatorial explosion would have quickly paralysed any truly general-purpose computational mechanisms. Recall that the main reason for this claim has been that because such computational mechanisms do not posess domain-specific constraints that help to narrow down the alternatives to a problem, the computational mechanisms are forced to evaluate each alternative in turn (Cosmides and Tooby, 1994b). Samuels (19998, 2000) points out, however, that whilst it is correct to dismiss (B)-type general-purpose computational mechanisms on the grounds of combinatorial explosion, since such computational mechanisms lack any domain-specific constraints that protect the computational mechanism against it, the same is not necessarily true of (A)-type general-purpose computational mechanisms. Once again, it appears that general-purpose computational mechanisms can possess specialised knowledge that helps constrain the alternative solutions to a problem if these computational mechanisms can access the relevant bodies of domain-specific knowledge. Hence, Samuel's library model of cognition cannot be ruled out on the basis of combinatorial explosion either.

To summarise Samuels (1998, 2000) claims: as it turns out, none of evolutionary psychologists' arguments are strong enough to support evolutionary psychology's double claim that, firstly, the mind is largely or even entirely composed of Darwinian modules and that, secondly, general-purpose computational mechanisms are ruled out on account of their evolutionary implausibility. Even conceding that the arguments of evolutionary psychology reveal an important trend in the evolutionary process, namely that it will lead to specialized cognitive structures, the arguments are not sufficient to exclude (A)-type general-purpose computational mechanisms from evolution. As Samuels correctly argues, the evolutionary trend to specialisation may be satisfied on the basis of innate and domain-specific bodies of knowledge.

However, the success of Samuels' arguments depends on the precise reading of evolutionary psychology. Even though I am sympathetic to Samuels' arguments, I believe that his reading of evolutionary psychology is tendentious and that, differently described, evolutionary psychology will not be equally affected by his arguments. On a first account, Samuels seems to be committed to what I have introduced earlier as the strong version of evolutionary psychology, which claims that the mind is entirely composed of both knowledge modules as well as computational modules. There are two observations that converge in assigning Samuels' reading of evolutionary psychology to the strong version of it:

First, Samuels' definition of evolutionary psychology as a view emphasising the importance of Darwinian modules in conjunction with his acceptance of knowledge modules in his library model of cognition, suggests that Samuels is strongly committed to a position that sees the mind as largely or even entirely composed of *both* computational modules *as well as* knowledge modules. For it would be inconsistent if Samuels allowed knowledge modules to play an important part in the case of his library model of cognition, but not so in the case of evolutionary psychology – since he accepts that knowledge modules may be the product of natural selection. The distinguishing feature between evolutionary psychology and his library model of cognition lies not in the fact that the former theory does not employ knowledge modules, but rather in the fact that evolutionary psychology puts a special emphasis on computational modules. Seeing that knowledge modules are necessary for the library model of cognition, and that computational modules are necessary for evolutionary psychology, Samuels has to interpret evolutionary psychology as claiming that the mind is largely or even entirely composed of both computational as well as knowledge modules.

Second, Samuels' arguments depend for their effectiveness on a version of evolutionary psychology that explicitly rejects general-purpose computational mechanisms. Otherwise his library model of cognition would not have the same effects against evolutionary psychology. The only version of evolutionary psychology that is capable of excluding general-purpose computational mechanisms is the strong version of it. For the weak version of evolutionary psychology requires only that the mind be largely composed of either knowledge modules or computational modules (where the – or – has to be read in a non-exclusive way). According to the weak version of evolutionary psychology, general-purpose computational mechanisms are allowed as long as they occur within a context of otherwise specialised cognition of either variety. Thus the weak version of evolutionary psychology does not reject general-purpose computational mechanisms.

It is already clear from the consideration outlined above, that if the weak version of evolutionary psychology turns out to be the proper target for criticism, Samuels' arguments do not affect this version of evolutionary psychology. Rather, his arguments could now be seen as lending further support to the weak version of evolutionary psychology. There are two observations that show why the weak version of evolutionary psychology should indeed be preferred:

On the one hand, as I already explained in chapter I, the weak version of evolutionary psychology has the advantage of formulating a substantial view of the mind that is less controversial than the strong one and that can therefore be accepted more easily by scientists working in the field. Its subjection to criticisms will show whether or not it is capable of offering a fruitful and consistent empirical programme that can be pursued. On the other hand, there are good grounds to think that the weak version of evolutionary psychology is the version that evolutionary psychologists themselves intended. One can find many passages in the work of evolutionary psychologists that clearly demonstrate that their preferred model of the mind includes both special-purpose computational mechanisms as well as bodies of domain-specific knowledge. Cosmides and Tooby (1994b, p. 94), for instance, write that:

> a domain-general evolved architecture is defined by what it lacks: It lacks any content either in the form of domain-specific knowledge or domain-specific procedures that can guide it towards the solution of an adaptive problem.

In this passage Cosmides and Tooby contrast their preferred domain-specific cognitive architecture with a domain-general cognitive architecture by showing what the latter lacks in comparison with the former, namely "domain-specific knowledge or domain-specific procedures". Again, I believe that it is reasonable to read the "or" as non-exclusive. General-purpose computational mechanisms can therefore be allowed into the picture as long as they occur within a cognitive architecture that is largely composed of either knowledge modules or computational modules (or indeed an admixture of both types of modules).

Both of these observations strongly militate against Samuels' suggestion of defining evolutionary psychology in a strong way. But once it is understood that the weak version of evolutionary psychology is to be preferred, Samuels' arguments for (A)-type general-purpose computational mechanisms can be used to strengthen the case for the weak version of evolutionary psychology.

3. Conclusion

The result of this chapter is that weak versions of evolutionary psychology defy all attempts at disproving them. While some of the objections call for small changes in evolutionary psychology, none of them do anything to call into question the consistency of the empirical research programme. Thus, weak versions of evolutionary psychology present a promising account of the mind that is not only able to integrate findings about the mind's modular structure, but is also able to accord general-purpose cognitive mechanisms a place within the mind's structure.

In order to prove that weak versions of evolutionary psychology are consistent, I evaluated first objections to evolutionary psychology's methodology and its reliance on the explanatory and predicative projects.

Objections to the explanatory project, on the one hand, pointed out that it looks completely gratuitous from the perspective of cognitive psychology. Since the explanatory project starts with cognitive capacities that are known independently of the evolutionary consideration, it would seem that the goal of cognitive psychology is already accomplished. A further insistence on the evolutionary function of cognitive capacities would not seem to be called for, and would, in fact, hinder progress in psychology. Against this conclusion, I showed that the insistence on evolutionary function indeed helps cognitive psychologists to understand better the intricate details of complex cognitive capacities.

Objections to the predicative project, on the other hand, were worried about the status of our interpretations of past environments. Some objections pointed out that these interpretations of past environments are too speculative, other objections undermined the interpretations by pointing to the effects that possible mismatches between past and modern environments could have on our knowledge. Against these conclusions, I showed that these worries are unfounded. Evolutionary psychologists have enough resources to clarify how our interpretations of past environments can be enriched in scientifically respectable ways. After thus showing that the objections to the methodology are groundless, I then moved on to the evaluation of the objections to basic concepts of evolutionary psychology.

Against Gould, who claims that evolutionary psychology is obsessed with adaptation, I showed that evolutionary psychologists are aware of possible alternative evolutionary explanations beside natural selection. Against Miller, who claims that evolutionary psychology neglects sexual selection, I claimed that evolutionary psychology's negligence does not constitute a grave problem and can be corrected easily. Against Karmiloff-Smith, who claims that the mind is an ontogenetic product of development, I claimed that evolutionary psychologists can develop counter-strategies against her interpretation of the developmental data that support her views.

In a final move I analysed Samuels' library model of cognition as a polemical alternative to strong versions of evolutionary psychology. While I am strongly sympathetic to most of Samuels' objections to strong versions of evolutionary psychology, I claimed that his objections do not affect weak versions of evolutionary psychology in similar ways. Weak versions of evolutionary psychology, as Samuels' library model of cognition, are able to accord general-purpose cognitive capacities a place within the mind's structure.

Additionally, the analysis of Samuels' library model of cognition showed that important questions remain about the evolutionary status of general-purpose cognitive capacities. Whether the postulation of such general-purpose cognitive capacities can be justified can only be decided on the basis of an evolutionary account of these cognitive capacities. The next chapter will elaborate on the evolutionary status of general-purpose cognitive capacities and will formulate a model that discusses conditions conducive to the evolution of specific general-purpose cognitive capacities in some non-human animals.

III. The Evolution of General-Purpose Cognitive Mechanisms

In the last chapter we saw that the weak version of evolutionary psychology is not affected by objections to it. This version of evolutionary psychology claims that the mind is composed largely of cognitive modules, and it does accord general-purpose cognitive mechanisms a place within the mind's structure. Additionally, the earlier analysis of Samuels' library model of cognition demonstrated that there remain important questions about the evolutionary status of general-purpose cognitive mechanisms. With evolutionary psychology thus intact, attention needs to shift from questions about individual cognitive modules to questions about the evolutionary status of general-purpose cognitive mechanisms.

This is because evolutionary psychology cannot be satisfied with the demonstration *that* general-purpose cognitive mechanisms are possible (as with Samuels' library model of cognition). Evolutionary psychology has to provide reasons that show *how* and under *what* conditions these general-purpose cognitive mechanisms could have evolved in some species: it needs to argue that an evolutionary analysis can be given of these general-purpose cognitive mechanism. Then, and only then can general-purpose cognitive mechanisms claim their rightful place in the mind's cognitive structure. It is to this task of an evolutionary analysis that the chapter turns next.

The chapter is structured as follows: Section 1. discusses different general-purpose cognitive mechanisms in order to bring out the explanatory target of the evolutionary analysis. Section 2. then introduces a distinction between an inner selective environment and an outer selective environment in order to distinguish the specific problems the mind faces because of its massive modularity from the problems it faces because of adaptive problems of its surroundings. The problems of the inner selective environment are the ones the mind faces because of the simultaneous activation of different cognitive modules. The section argues that these problems of the inner selective environment amount to evolutionarily relevant problems that the mind had to solve in its evolution. Section 3. introduces and evaluates a model that evolutionary psychologists have recently put forward that tries to solve problems of the inner selective environment by ways of special-purpose cognitive mechanisms. The evaluation will show that this model is far from convincing and thus that general-purpose cognitive mechanisms are the better alternative to solve these specific problems. Section 4. starts the positive evolutionary analysis of general-purpose cognitive mechanisms by discussing the idea that unpredictability in the environment constitutes the proper domain of general-purpose cognitive mechanisms in general. Section 3.5, based on the insights from the previous section, develops an evolutionary account of the correct type of general-purpose cognitive mechanism by analysing and reworking a model recently proposed in the literature.

I must, however, issue an important caveat. The proposal I am about to outline is speculative. This is among other things due to the state of current research in animal psychology. While the postulation of modules to explain animal behaviour is well entrenched within animal psychology, issues about the interactions of modules have not yet received appropriate attention. Sara Shettleworth has recently highlighted this situation in animal psychology in writing that "ideas about how such hierarchies [modular hierarchies], or other ways of combining information, work are not very well developed." (2000, p. 55). Thus my proposal will use whatever clues it finds useful in fleshing out the details of the evolutionary account. With this caveat in mind, let me proceed with the discussion of the chapter.

1. The explanatory target of the evolutionary analysis

It is a reasonable condition on any form of analysis that it should affirm its proper explanatory target at the outset. Thus the evolutionary analysis of general-purpose cognitive mechanisms needs to affirm the type of general-purpose cognitive mechanism that acts as its proper explanatory target. This is done initially by distinguishing different forms of general-purpose cognitive mechanisms from each other. Samuels (1998), as we saw in the last chapter, proposes a distinction between different forms of general-purpose cognitive mechanisms that I mention here again:

(C) General-purpose computational mechanisms that are not special-purpose, i.e. ones that are not dedicated to solving problems in a single specific domain.

(D) General-purpose computational mechanisms that are not special-purpose – *and do not benefit from any innate, domain-specific knowledge or innate, special-purpose mechanisms.*

While his distinction is certainly valuable in clarifying the central debates surrounding evolutionary psychology, it is still not precise enough to pinpoint exactly the explanatory target of the evolutionary analysis. To make this plain I adduce an initial observation.

The initial observation is the fact that not all types of general-purpose cognitive mechanisms are equally problematic from an evolutionary psychology's perspective. Appearances notwithstanding, evolutionary psychologists actually believe that there *are* general-purpose cognitive mechanisms in the mind's cognitive structure, for instance, trial-and-error procedures and instrumental learning. Evidence for such a belief on the part of evolutionary psychologists is to be found in typical passages such as the following:

> we have made some serious and extended arguments about why many more cognitive mechanisms will turn out to include domain-specific features *alongside whatever domain-general principles of operation they have.* (Tooby and Cosmides, 1998, p. 198, emphasis added)

Clearly then, evolutionary psychologists concede that the mind contains general-purpose cognitive "principles of operation" that work alongside special-purpose ones. However, if that is the actual situation, the real controversy between evolutionary psychologists and their opponents does not concentrate on these uncontroversial general-purpose cognitive mechanisms – since both sides accept them anyway. The real controversy is about different types of general-purpose cognitive mechanisms; general-purpose cognitive capacities that would likely raise suspicion from the camp of evolutionary psychology. What are these types of general-purpose cognitive mechanisms? The answer to this question is found as soon as we recall those features of general-purpose cognitive mechanisms that made them suspicious for evolutionary psychologists in the beginning.

Some evolutionary scientists (Barkow, Cosmides and Tooby, 1992; Shettleworth, 1998) have argued that the content-independent nature of general-purpose cognitive mechanisms, their slow learning rate and their high propensity for error do not permit the production of sophisticated behaviour necessary for solving adaptive problems. It is cognitive modules that provide the necessary behaviour because they at least possess content-dependent principles that are tailor-made for the solution of adaptive problems. We can deduce from this consideration the types of general-purpose cognitive mechanisms that would elicit controversy from these evolutionary scientists. It is the general-purpose cognitive mechanisms that can profit somehow from content-dependent special-purpose cognitive capacities in order to produce the sophisticated behaviour necessary for solving adaptive problems.

Framed like this, these latter general-purpose cognitive mechanisms belong to Samuels' general-purpose cognitive mechanisms of type A of which his library model of cognition provides an example.

Samuels' model employs a central general-purpose cognitive mechanism that profits from the content-dependent operations of knowledge modules by accessing them in its reasoning. Thus this central general-purpose cognitive mechanism enhances the mind's problem solving abilities because it sits so to speak on top of the mind's massive modular machinery with which it conjointly operates to solve the mind's day to day businesses.

We are now in a position to refine Samuels' distinction above by stressing the structural relations that hold between general-purpose cognitive mechanisms and cognitive modules in the following manner:

(Type 1) General-purpose cognitive mechanisms that do *not* have access to a massive modular cognitive structure (=GPCM 1).

(Type 2) General-purpose cognitive mechanisms that have access to a massive modular cognitive architecture.

We can localise these two types of general-purpose cognitive mechanisms in a simple structural chart of the mind's cognitive architecture:

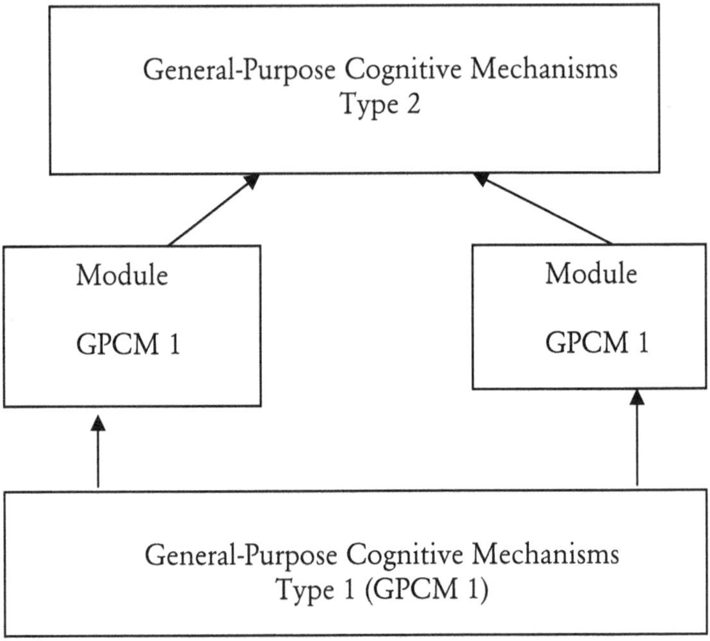

As we see from the picture above there are two locations where general-purpose cognitive mechanisms of type 1 can be situated.

On the one hand, these cognitive mechanisms can be localised at the bottom of the cognitive architecture where perceptual information reaches the mind and is routed to various modules inside the architecture where information is further processed. The bottom of the picture symbolises the organisms' general sensorium that collects information coming from the perceptual organs before it passes it along.

On the other hand, general-purpose cognitive mechanisms of type 1 may be situated inside cognitive modules and could function there alongside whatever other content-dependent principles the modules possess. Moreover, it is possible that different cognitive modules implement identical general-purpose cognitive mechanisms of type 1 since each implementation of these general mechanisms would be separated from each other by the cognitive modules' informational encapsulation from other modules. No conflict between the nature of cognitive modules and the nature of general-purpose cognitive mechanisms of type 1 needs to arise as long as the latter cognitive mechanisms obey the module's informational encapsulation.

But as we have already observed, general-purpose cognitive mechanisms of type 1 are not the most interesting ones from an evolutionary perspective. The most interesting ones, by contrast, are those general-purpose cognitive mechanisms that sit on top of the massive modular system and benefit from cognitive modules by accessing their content-dependent principles: we need to focus on general-purpose cognitive mechanisms of type 2.

2. Inner Selective Environment and Outer Selective Environment

Now that we know which general-purpose cognitive mechanisms are the explanatory target of the evolutionary analysis, we need to find out something about the kind of problems these cognitive mechanisms were likely to solve. We can approach this issue in the following way.

Evolutionary psychologists have so far concentrated their research on individual cognitive modules that solve specific adaptive problems that originate from the outside environment. By contrast, they have not systematically analysed problems stemming from the interaction between cognitive modules within the mind. It will help our discussion if we have terms that explicitly distinguish these two different kinds of problems. To this end, I introduce the term *inner selective environment* to refer exclusively to problems stemming from the interaction of modules, and I distinguish this from an *outer selective environment* that refers to problems originating outside the organism in its physical and social environment.

Now the concept of an outer selective environment is already familiar, albeit not by this name, from our discussion in chapter I. The outer selective environment is closely related to the environment of evolutionary adaptedness. It includes the totality of selection pressures from adaptive problems in the outer ecological environment of a particular species. Evolutionary psychologists' core arguments introduced in chapter I have shown why on the basis of the outer selective environment, specialised cognitive structures are generally better suited to solve adaptive problems on account of their speed, reliability and efficiency. Nothing more needs to be said about the outer selective environment here.

The inner selective environment, on the other hand, is not directly concerned with outer adaptive problems as such. Consequently, this environment does not provide arguments for a modular mind, since it already assumes that such a mind is actually in place. Rather, the inner selective environment's main interest lies in outlining the different types of interactions between modules, and in understanding what kinds of cognitive mechanisms are necessary to solve modular interaction. This is not to say that the consideration based on the inner selective environment may not concern itself with situations the mind has to face from outside. But such situations are of interest only in so far as they give rise to problems of modular interaction within the mind.

We can better appreciate the kinds of problems that the inner selective environment poses for the mind by inquiring why the inner selective environment emerged in the first place. One important insight that such an inquiry discloses is that organisms, at any given moment of their life, are exposed to complex situations that involve the presence of multitudinous information directly relevant to the survival or reproduction of the organism. Real situations do not typically involve just one specific type of information that can be taken care of by a particular module. Instead, real life situations will generally involve the simultaneous occurrence of different types of information that need to be taken care of by different modules at the same time.

Given the plausible assumption (Gigerenzer, 1997) that each module has characteristic triggering algorithms that parse the accessible perceptual or conceptual information at each given instant to see if it finds matching input, the complex situations will inevitably trigger several modules into action at the same time. Once modules have thus been activated simultaneously, they will enter into competition with each other. The problem for the mind at this point is to find ways to integrate, coordinate, or combine these modules in the most effective and beneficial ways possible for the organism. In other words, the mind needs to impose some sort of priority ordering on the modules.

The idea of the competition of modules for behavioural output can be brought into even sharper focus by an analogy. In analogy to the limiting resources of the outside ecological environment, e.g. food, rivals, and mates, the internal modular "environment" has its own limiting resource for the control of behavioural pathways. An organism can only do one thing at a time, so it must decide which module should influence its behaviour. Consequently, as with organisms competing over environmental resources, the modules enter into competition over access to the motor systems of the organism. And it will not be the case that all modules have equal access to these motor programmes. Again, as in the case of competing organisms, modules will have differential success in accessing the motor programmes. Some modules, for instance, will be better equipped than others to win this competition for the limiting resource and as a consequence have greater access to behavioural pathways. The outcome of the competition, i.e. which module gains precedence over which, has obvious ramifications for the reproductive success of the organism. If the wrong module wins, or if a combination of modules leads the organism astray, it could mean death. But if, on the other hand, the right module wins, or if a particular combination of modules enhances performance, the organism may improve its reproductive success and it may pass on this solution to the next generation either by genes or memes.

The observations above, though, may easily lead to an incorrect picture of the inner selective environment. It may suggest that modules could evolve prior to and independently on their being integrated, coordinated, or combined in some ways. Only at a later stage of evolution would the modules then enter into a specific pattern of cooperation with each other. However, such a picture is highly implausible for the following reasons.

Natural selection can only act upon observable behaviour, which is caused by the whole adaptive structure of an organism. If the behaviour is reproductively successful, the whole adaptive structure that is implicated in the production of the behaviour will be represented in the next generation. But if the behaviour is a failure reproductively, the whole adaptive structure of this particular organism may perish and thus will not be represented in the next generation. Natural selection cannot, so to speak, dodge behaviour to reach inside the adaptive structure and alter just one specific module in isolation of all the others. It cannot disregard the influences these other modules have on the behaviour.

Certainly then, for modules to evolve at all, they must make a positive contribution to the behaviour of the whole organism. But this implies that a module must already cooperate with whatever other modules are present in the mind to result in reproductively successful behaviour, and thus be *selected for*. Given this consideration, it is wrong to think that single modules could evolve in complete isolation from all others by somehow circumventing issues of cooperation. The integration, coordination, or combination of modules is a problem the mind faces from its very beginnings, and it is not something that suddenly pops up as soon as the mind reaches a critical mass of modular complexity.

So the mind needs some forms of cooperation among modules. But might not selection pressures arise on the basis of a given form of cooperation of the mind's components and propel the mind forward into improved levels of cooperation among modules?

Natural selection would thus drive the mind from one level of co-operation to another improved level of cooperation. Perhaps these improved levels of cooperation would as a consequence even improve the workings of individual modules (i.e. by freeing these modules from constraints imposed on them by the initial cooperation with other modules and allowing them thereby to develop fully their evolutionary potential). My concept of the inner selective environment is designed to include such consideration.

Depending on the type and inherent structure of a complex situation the organism faces, it may develop different solutions to the problems of modular interaction. I shall briefly outline two different solutions to modular interactions in order to clarify further the issues the mind has to confront, before moving on to a more detailed discussion.

Conflict Mitigating Strategies

Animals often encounter situations that lead to conflicting demands on their cognitive structure. A concrete example of such a situation is one in which a food-deprived animal both perceives a rich food source and a predator lurking in the vicinity of the food source. The perceived situation will inevitably activate both food approach behaviour (due to a foraging module) and predator evasion behaviour (due to a predator detection module). In the absence of conflict mitigating strategies, the animal is prone to fatal errors. The situation may either arrest the animal's ability to respond in case the animal wavers between the two options of approaching the food source or running away from it, or the animal may give priority to the wrong module in the context of the situation. Both options may easily prove fatal to the animal. In order to shield the organism from the fatal consequences of wrong decisions, the mind will resort to conflict mitigating strategies that allow it to solve the conflict between cognitive modules by automatically favouring the more important cognitive module in the context, and by automatically suppressing all the others.

If, in our example, the animal is near starvation, it will make sense to suppress the consequences of the predator detection module even if the danger is highly salient for the animal in the context. A risk prone strategy may benefit the animal better in the given context.

Integration, Coordination, and Combining Strategies

Animals also encounter types of situations that simultaneously activate cognitive modules, but whose solution requires that the activated modules be somehow integrated, coordinated, or combined with each other. These solutions are more sophisticated than the conflict mitigating strategies because they produce solutions that go beyond merely favouring particular modules by suppressing all others. These new solutions can turn the simultaneous activation of modules to the organism's advantage by integrating, coordinating, or combining modules to derive adaptive responses.

A concrete example of the use of these solutions is found in animals that need to negotiate their social status in hierarchically stratified social groups (Russon, Bard, and Taylor Parker, 1996; Tomasello and Call, 1997). The social position an organism occupies in a social group is strongly correlated with its survival or reproductive success. It is very probable that the successful attainment of a particular social position in a group requires the simultaneous activation of specialised cognitive equipment – such as cognitive programmes specialised in the recognition of individual group members, memory systems concerned with the accurate retrieval of information about past behaviour of particular group-members, cognitive programmes specialised in dealing with the social relation between group-members, and cognitive programmes dealing with the behaviour of physical objects (in case the animal must resort to violent action to dethrone other group-members from their social position). The success of the animal's behaviour in attaining its goal depends on the kinds of strategies the animal uses when it integrates the different cognitive modules with each other.

On any account, in one way or another, organisms faced with situations that simultaneously activate different cognitive modules need to impose strategies upon them to secure beneficial behaviour in those situations. The strategies they can use to solve adaptive problems consist in either conflict mitigating strategies or strategies of integrating, coordinating, or combining modules in correct configurations. As we found earlier, natural selection primarily acts on the behaviour that such strategies produce because it *cannot* act upon the performance of individual cognitive modules in isolation from all others. How the organism behaves when confronted by its simultaneously activated modules thus has a major impact on its survival and reproductive success.

3. Evolutionary psychology and Inner Selective Environment

Until very recently, evolutionary psychologists have neglected issues similar to the ones raised by the inner selective environment. A natural question, therefore, is whether evolutionary psychologists possess arguments that can deal successfully with the problems of the inner selective environment. More specifically, can evolutionary psychologists use their arguments in favour of specialised solutions in the domain of the inner selective environment as well in a way that could undermine general-purpose cognitive mechanisms? This question needs to be investigated next.

Evolutionary psychology's argument from statistical information can act as a guide into this investigation. As we have seen, Cosmides and Tooby (1994b), for instance, argue that most of the information that is necessary for solving adaptive problems will not be observable by the individual organism in its lifetime since the recurrent structure of the information will emerge only over long periods of time. Natural selection, however, can take advantage of this information by building innate cognitive devices that are able to profit from this information in solving the problems. As a consequence of natural selection, the cognitive devices internalise the relevant task demands of the domain into their design.[32]

The question is whether a similar consideration is applicable to the problems of the inner selective environment. Is it possible that evolution could have exploited statistically recurring information in the environment in order to develop fixed solutions that orchestrate cognitive modules to effect beneficial responses?

Cosmides and Tooby's (2000) new theory of emotions develops a proposal that aims to show how special-purpose solutions can take account of the problems of modular interaction. Their proposal focuses more on the integration strategies than on the conflict mitigateing strategies, but actually uses both strategies to solve these specific problems. In what follows I shall briefly outline their proposal since it constitutes the best recent attempt of evolutionary psychologists to deal with problems similar to the ones originating with the inner selective environment.

The basic idea underlying Cosmides and Tooby's theory of emotions is the idea that specific adaptive problems require that the mind develops cognitive programmes, so called "superordinate" programmes, that combine different cognitive modules into the right kind of configuration at the right time. According to Cosmides and Tobby (1990, p. 2), "many adaptive problems are best solved by the simultaneous activation of many different *components* of the cognitive architecture, such that each component assumes one of several alternative states (e.g. predator avoidance may require simultaneous shifts in both heart rate and auditory acuity)".

In Cosmides and Tooby's proposal, a "superordinate" cognitive programme is realised as an emotion, which they define thus:

32 The concept of *internalisation* is a well-entrenched scientific concept that is used especially in studies of the visual system. For a theoretical discussion of internalisation and some of its empirical applications see especially Shepard (1984, 2001). I trust that the concept of internalising outside regularities in the design features of tools is intuitively understood here.

> an emotion is a superordinate program whose function is to direct the activities and interactions of the subprograms [...] (Cosmides and Tooby, 2000, pp. 2-3)

Cosmides and Tooby argue further that emotions are special-purpose cognitive programmes that evolved specifically to deal with the competition between modules highlighted by the inner selective environment. The reasoning they use to support their case about emotions is an example of their argument from statistical information. As they write:

> Fighting, falling in love, escaping predators, confronting sexual infidelity, experiencing a failure-driven loss in status, responding to the death of a family member (and so on) each involved conditions, contingencies, situations, or event-types that recurred innumerable times in hominid evolutionary history. Repeated encounters with each kind of situation selected for adaptations that guided information-processing, behavior and the body adaptively through the clusters of conditions, demands, and contingencies that characterized that particular class of situations. This could be accomplished by engineering superordinate programs, each of which jointly mobilizes a subset of the psychological architecture's other programs in a particular configuration. Each configuration would be selected to deploy computational and physiological mechanisms in a way that, when averaged over individuals and generations, would have led to the most fitness-promoting subsequent lifetime outcome given that ancestral situation-type. (Cosmides and Tooby, 2000, p. 2)

The rationale for Cosmides and Tooby's claim that emotions are special-purpose cognitive mechanisms is clear. Our ancestors' repeated encounters with specific situations selected for special-purpose cognitive programmes that deal with those situations. The complexities involved in these specific situations mandated a complementary complexity in the cognitive programmes leading to programmes that were capable of integrating various components of the mind into correct configurations, which could solve the specific problems. These special-purpose cognitive programmes are realised in the human mind by emotions. Particular emotions such as *anger* or *sexual jealousy* are therefore associated with particular constellations of cognitive components.

Emotions, as all other forms of cognitive modules, possess characteristic triggering algorithms that parse incoming information for relevant clues that will activate the module (Cosmides and Tooby, 2000). In the case of emotions, these triggering algorithms check whether the relevant type of situations is present. If the triggering algorithm detects a situation, it activates the relevant emotion programme. Once activated, the emotion programme produces a signal that summons the cognitive components into the specific constellation needed to solve problems originating with the situation. The organism does not have to learn about the particular constellation of cognitive components because it has already been pre-configured by evolution.

Emotions differ from other types of cognitive modules only in the light of their higher-order complexity that includes mechanisms to integrate or combine cognitive modules.

Cosmides and Tooby's (2000) analysis of emotions has wider ramifications for their massively modular mind. The massively modular mind receives a strongly hierarchical structure with the inclusion of emotions. Not only does the mind possess first-order cognitive modules that take care of specific problems, it also contains emotions that integrate first-order cognitive modules, which thus constitute second-order cognitive modules.

The outcome of the emotion analysis seems to show that evolutionary psychology indeed possesses arguments to address the specific problems of the inner selective environment in special-purpose ways. If sound, these arguments could undermine the alternative general-purpose cognitive account of these problems of the inner selective environment. The important question is therefore whether evolutionary psychology's arguments are correct. I do not believe they are and there are two reasons for my claim.

Objections 1

One objection to Cosmides and Tooby's proposal starts from their proposed analysis of emotion. While their analysis of emotions may account for *some* modular integration, it is not sufficient to account for the whole range of problems the inner selective environment poses. To recognise this, one only needs to ask the following question: What happens, for example, if two or more emotions enter into conflict with each other? What mechanisms can the mind bring to bear on these situations?

This competition or conflict between emotions looks like just another instance of the simultaneous activation of cognitive modules that initially mandated the introduction of second-order cognitive modules. This time though it is not the simultaneous activation of first-order cognitive modules that creates the problem, but rather the simultaneous activation of second-order cognitive modules, i.e. emotions. Therefore, the mind is faced again with the same problem of finding ways to integrate *emotions* in order to benefit the organism. Although Cosmides and Tooby do not systematically address this new problem in their paper, one can make informed conjectures about their probable response.

Cosmides and Tooby might suppose that the simultaneous activation of particular emotions would have occurred innumerable times in evolutionary history, and so would have given rise to an adaptive problem of its own. This would have constituted a sufficiently complex situation, which our ancestors would have had to solve somehow. Moreover, evolutionary psychologists could say that these situations involved enough recurrent patterns of contingencies and conditions which natural selection could then have exploited in developing special-purpose cognitive mechanisms for dealing with conflicting emotions. On analogy with Cosmides and Tooby's concept of *emotion*, we may call these new third-order special-purpose programmes *meta-emotions.* Meta-emotions would probably solve their specific problems in analogous ways – mobilising the right constellation of emotions, deactivating some, activating others.

Again, the inclusion of meta-emotions would have ramifications for the massively modular mind which would receive an even stronger hierarchical structure: the mind would contain first-order cognitive modules, second-order emotional programmes, and third-order meta-emotional programmes.[33]

Maybe this is what evolutionary psychologists should argue for. But it is not obvious how likely or unlikely this is from an abstract description of such emotional conflict situations. In any case, evolutionary psychologists have not provided reasons to show what happens if a competition or conflict arises between particular meta-emotions. What can the mind do then? Is it possible that the repeated encounters with these new situations could have selected for meta-meta-emotions, etc? I am not saying that these situations are likely to occur, but evolutionary psychologists have not given any reasons to show that they will not occur in principle. This consideration, however, may support worries about an infinite regress. Is there a way out of the predicament?

An obvious exit from these worries is to postulate general-purpose cognitive mechanisms of type 2 that can access the various conflicting emotions and find the most beneficial solution to specific problems in the light of the situational context. Freed by the constraints typical of cognitive modules, the general-purpose cognitive mechanisms of type 2 are free to negotiate a new solution for each new situation. In any case, this is not a solution, which is open to Cosmides and Tooby since their aim is to argue for special-purpose solutions to these types of problems.

Interestingly, Cosmides and Tooby (2000) hint at a solution to the problem of emotional conflict that circumvents the postulation of general-purpose mechanisms without falling into the trap of an infinite regress. Their solution is very similar to the fixed priority orderings I discussed earlier – and thus supplements the function of

33 Sperber (1994) argues along similar lines. He argues that a hierarchy of domain-specific and innate computational mechanisms might effect the integration of different modular outputs, which is typically regarded as the function of our human reasoning abilities. Modules higher up the hierarchy are more complex than other modules beneath them because they would be responsible for more complicated integrations of the information channelled through the system.

integration that emotions perform. Fixed priority orderings, as we have seen, are formed and shaped into the cognitive design of organisms by natural selection. The relevant passage of Cosmides and Tooby reads:

> Depending on the relative importance of the situations and the reliability of the cues, the prioritizing algorithms decide which emotion modes to activate and deactivate, and to what degree. Natural selection would have sorted emotions based on the average importance of the consequences stemming from each, and the extent to which joint activation was mutually incompatible (or facilitating). (Cosmides and Tooby, 2000, p. 12)

The fixed priority orderings between emotions, or the "prioritizing algorithms" as Cosmides and Tooby call them, constitute an ordering of emotions, which is sorted once and for all *by natural selection* according to the (average) importance of the emotion's consequences to the organism *and* the compatibility of their joint activation. As an evolutionary product the priority ordering is the end-result of a long process in which natural selection had to find the most advantageous relative ordering possible for the organism. Whenever a conflict between emotions occurs, the mind will solve the conflict by automatically favouring the more important one, and by automatically suppressing all the others.[34]

34 There is an ambiguity in the Cosmides and Tooby (2000) passage cited above. Note that Cosmides and Tooby write about *prioritizing algorithms* that decide which emotions are activated and which are deactivated. Now in common parlance, the term *algorithm* denotes a computational *rule or procedure*, an operation that manipulates *explicitly represented computational states*. So, reading the passage, one might get the impression that Cosmides and Tooby suggest the postulation of a species of computational mechanism. But as the context of the whole passage demonstrates, this cannot be correct. In the rest of the passage, Cosmides and Tooby argue for a fixed priority ordering that natural selection has wired into the human mind. So now it sounds as though Cosmides and Tooby argue for the view that natural selection has found a hard-wired solution by instantiating it *(implicitly)* in the cognitive architecture. I think that it is the latter interpretation that captures Cosmides and Tooby's solution. If, on the contrary, it is the algorithmic interpretation they prefer, then I do not see how they can exclude the view that it is general-purpose

It needs to be said, though, that Cosmides and Tooby's solution is, at present, nothing more than a plausible suggestion. It has its clear limitations. Among other things, its intuitive plausibility rests completely on very specific type of ordering; an ordering in which each item is assigned a definite position on a rank that separates it from any other item. We can consider a possible event to bring out the particular nature of the ordering even clearer.

For what would happen if, for instance, two emotions conflicted with each other, but were located on the same rank of the ordering? With no emotion being more important than the other, the mind cannot solve this conflict by referring back to its fixed priority ordering – since it cannot arrive at a decision in the case the two emotions are equally important. To call on a random generator to decide in those circumstances is not a good idea, for the solution of these conflicts may very well affect the organism in important ways.[35]

It would appear, therefore, that a fixed priority ordering as Cosmides and Tooby envisage could only work successfully in those circumstances in which a very special type of ordering existed. In the case such a special type of ordering does not exist, different types of mechanisms are needed to solve the conflicts. These alternative cognitive mechanisms could not depend on fixed solutions because the required orderings for fixed solutions are absent. They will thus have to employ more flexible tools in their reasoning. But once we grant the flexibility of the mechanisms, we are back with general-purpose cognitive mechanisms of type 2 as the prime candidate that

cognitive mechanisms that realize the relevant algorithms rather than special-purpose ones.

35 Cosmides and Tooby would have to say that the actual solution of such a conflict among emotions has important consequences for the organism's reproductive success, since they themselves state that natural selection could have instantiated an ordering on the basis of the relative importance of the emotion's consequences. The whole point of the passage is to claim that conflict situations are evolutionarily important situations the organism needs to face.

can realise it. Confronted by emotional conflicts, the mind would have to access the conflicting emotions and reason out a solution that is the most advantageous for the organism in the actual context of the situations it finds itself in.

Nothing in Cosmides and Tooby's (2000) arguments so far allows them to rule out general-purpose cognitive mechanisms from an evolutionary account. The reasons I have extrapolated from their arguments in their paper are not sufficient to support their case for special-purpose cognition. To this end, Cosmides and Tooby would have to make a clearer and more convincing case for why we should expect the specific types of orderings to occur that could support their reasoning.

They could try to marshal the theoretical arguments of evolutionary psychology introduced in chapter I against general-purpose cognitive mechanisms. But as we saw in chapter II, these arguments are not strong enough to exclude general-purpose cognitive mechanisms of type 2 either – see Samuels' library model of cognition.

Consequently, general-purpose cognitive mechanisms of type 2 are not yet ruled out from an evolutionary perspective.

Objection 2

Another objection to Cosmides and Tooby is connected with the computational nature of emotions. Cosmides and Tooby define *emotions* as *superordinate programs* whose function is to supervise the interactions of other sub-programmes of the mind. Emotions are thus a species of computational mechanisms that were formed by natural selection to solve specific adaptive problems. The question is whether Cosmides and Tooby's arguments actually favour *special-purpose cognitive mechanisms* over *general-purpose* cognitive ones. They do not, and we already know why.

When discussing the various objections levelled against evolutionary psychology in chapter II, I presented the reasons why Samuels (1998, 2000) believes that evolutionary psychologists' theoretical arguments are not strong enough to exclude general-purpose cognitive mechanisms. While he agrees that the theoretical arguments show that the mind contains large amounts of specialised cognitive structures, he maintains that they could not show that the required specialisation is due to special-purpose computational mechanisms rather than to bodies of domain-specific knowledge.

Precisely the same kind of consideration applies to Cosmides and Tooby's (2000) proposal about emotions. Samuels' library model of cognition can account for the same types of situations Cosmides and Tooby adduce in support of their proposal just in the case the repeated encounters with the assumed types of situations would have selected for bodies of domain-specific knowledge instead of special-purpose computational mechanisms.

We already know that one requirement for the selection of specialised cognitive content is the presence of recurrent information in a domain. Cosmides and Tooby demonstrate in their argument that such recurrent information is to be expected since "each [type of situation] involved conditions, contingencies, situations, or event-types that recurred innumerable times in hominid evolutionary history." (2000, p. 2) So the requirement for specialisation is satisfied. Given the recurrent information, however, the library model of cognition can provide an alternative interpretation of the relevant situation. Viewed from the perspective of the library model of cognition, the mind is more likely to develop *bodies of domain-specific knowledge* that specify which other components of the mind are to be jointly mobilised in a particular configuration to solve specific adaptive problems. These bodies of domain-specific knowledge, however, are manipulated by general-purpose cognitive mechanisms of type 2. Moreover, because these general-purpose cognitive mechanisms would have access to different bodies of specialised knowledge they would not be condemned to treat all problems uniformly. In the light of Cosmides and Tooby's views about emotions and their implication for the structure of the massively modular mind, the library model of cognition can also be easily modified to include a set of *superordinate bodies of specialised knowledge.*

If this consideration is sound, the library model of cognition provides an alternative analysis of emotions. Whereas evolutionary psychology treats emotions as a species of *special-purpose computational mechanisms*, the library model of cognition, on the contrary, treats them as a species of *bodies of domain-specific knowledge.* I do not know whether there is evidence to decide between these two theories, but it is surely an interesting issue that is worth pursuing further on an empirical basis. In any case, since the library model of cognition accounts for exactly the same kind of problems as evolutionary psychology, it cannot be ruled out on account of those problems alone.[36]

Consequently, general-purpose cognitive mechanisms are once again not ruled out from an evolutionary perspective.

Concluding this section we can observe that Cosmides and Tooby (2000) have started to take problems originating with the inner selective environment seriously. Consistent with their general views about evolutionary psychology, they argued for special-purpose solutions to the problems of modular interaction. However, none of their arguments makes a convincing case for their special-purpose solutions. More specifically, the arguments are not capable of excluding general-purpose cognitive mechanisms of type 2 from the mind's structure. Firstly, I gave reasons to show how their special-purpose solutions to emotional conflict depend on unsubstantiated assumptions about particular types of orderings between emotions. Secondly, I argued that their definition of emotions as special-purpose computational mechanisms (superordinate pro–grammes) is not justified solely on the basis of assumed recurrent information in relevant situations. Cosmides and Tooby's definition

36 A further important point is that the library model of cognition can easily handle issues arising from conflicts between emotions. We saw in *Criticism 1* how emotional conflicts may, under particular situations (i.e. absence of specific orderings), lead to problems for Cosmides and Tooby's (2000) proposal. Since the library model of cognition already uses general-purpose cognitive mechanisms, it might as well use them to solve conflicts that arise between emotions.

of emotions in terms of special-purpose computational mechanisms portrays their unjustified assumption that evolution's tendency to specialisation must always issue in special-purpose computational mechanisms rather than in bodies of domain-specific knowledge. As a consequence of the arguments' failings, general-purpose cognitive mechanisms of type 2 still constitute viable cognitive options that need further evolutionary clarification. In the next section I will therefore start my investigation into the evolutionary grounds of these cognitive mechanisms that may have caused evolution to develop such cognitive mechanisms.

4. Evolutionary significance of different information

We already know from our discussions in chapter I that the structure of information is a crucial moment of evolutionary analyses of cognition. Evolutionary psychologists, for instance, develop their arguments for cognitive modules by highlighting those domains where the relevant structure of information shows high levels of predictability over generations. So if we are interested in an evolutionary analysis of general-purpose cognitive mechanisms, our initial move consists in clarifying the type of information to which general-purpose cognitive mechanisms respond. In the following, I want to do this by contrasting the type of information relevant to the evolution of general-purpose cognitive mechanisms with the types of information relevant to the evolution of special-purpose cognitive mechanisms.

A good way of specifying the different types of information the mind is exposed to, is by briefly re-examining once more one of the core arguments evolutionary psychologists use to support cognitive modules: the argument of statistical information from chapter I. A quick reminder of the essentials of this argument is in order here. Organisms are confronted by particular kinds of situations in evolution that involve conditions and contingencies the organism cannot observe in its lifetime. Yet many of these situations would have possessed sufficiently recurrent patterns of information, i.e. statistical regularities, that repeated encounters with each of them over evolutionary time would have selected for specialised cognitive adaptations. The success of evolutionary psychologists' arguments, as we said, depends on the fact that highly predictive patterns of information and regularities actually occur in the environment which organisms do not have access to on the basis of their general learning mechanisms. If such patterns indeed occur in the environment, natural selection will generally select for innate specialised cognitive structures.

What happens, though, if the environmental information does not manifest recurring patterns of this sort? Then the development of hard-wired or pre-set solutions may turn out to be grossly maladaptive for dealing with such situations. This is because organisms with hard-wired or pre-set solutions would not be able to cope with the shifting situations, and consequently, could not produce the required adaptive responses the situations demand.

A deeper analysis of the evolutionary psychologists' argument discloses two very different categories of predictable and recurrent patterns of information involved in certain situations of the environment.

Category 1:
Some types of situations involve conditions and contingencies that are fixed over the course of time and do not change significantly from one generation to the next. While organisms are not able to detect these regularities, evolution exploits these predictable regularities to fashion cognitive modules that incorporate the relevant task demands that are needed to solve problems in these domains. The environmental regularities are therefore stable and we can call these regularities *constant in the strong sense.*

A concrete example will help to clarify these situations. Natural light typically emanates in our environment from a light source above. This is a regularity that is constant in the strong sense and that holds over evolutionary time. Evolution has seized upon these regularities and internalised them into the design of the visual system, and the visual system uses its design to interpret its visual stimulus accordingly. As a consequence, organisms do not need to learn about such regularities (Shepard, 1984, 2001). Other instances of evolution seizing on regularities in the strong sense to build specialised structures are fixed action patterns that help organisms face specific problems by instinctive responses (Tinbergen, 1951). The laughing gull chick, for example, possesses programmes that let it obtain regurgitated food from its parents whenever it pecks the red dot at the end of its parents' beak. Seizing on the constant regularity between the presence of parents and access to food, evolution has given organisms a fixed-action pattern that allows the chick to get food without learning.

Category 2:
Other types of situations involve conditions and contingencies that are not fixed over the course of time and can change somewhat from one generation to the next. However, these changing regularities can change only within certain circumscribed limits. The variance of the changing conditions is highly limited. Thus it may be that predictable limits exist that constrain the range of variation that parameters undergo from one generation to the next. These constraints on the variance of the changing parameters may be what is required to retain some predictability of the contingencies involved in these specific situations. So while the *particular* instantiation of a regularity in a given context cannot be predicted, the *range of variance* of the parameters may be predict*able* over the course of time. Evolution is still capable of exploiting these kinds of regularities and to fashion cognitive modules that incorporate the relevant information that is needed to solve problems in this domain. We can call these environmental regularities *constant in the weak sense*.

While evolution is able to seize on the predictable features of regularities in the weak sense to build cognitive structures, certain forms of learning will be required to adjust or fine-tune these modules to the prevailing distribution of values in the ontogenetic environment of the organism. A paradigmatic example of such an evolved cognitive structure is a Chomsky-style language module (Chomsky, 1988). Evolution cannot predict the details of the particular grammar spoken in the language community, and will therefore have to be supported by learning mechanisms. But it is still able to determine the principles and parameters every possible human language must obey (the Universal Grammar of Chomskian linguistics). Consequently, humans possess a language module dedicated to processing linguistic input even though the particular grammars spoken in any given community could not have been foreseen by evolution. Another example of this involves bird songs. As it turns out, certain bird species are biologically prepared to learn quickly about the specific song-patterns in their vicinity or the stellar constellations they use for navigation, even though the local instantiations of these regularities may vary considerably across time and space (Gould and Gould, 1994; Marler, 1970).

The structure of information of category 1 and category 2 is the one evolutionary psychologists use in their arguments. Those informational structures demonstrate sufficient levels of predictability to allow evolution to seize upon the regularities and build specialised cognitive capacities that exploit them. However, the question arises whether these two categories exhaust the entire range of evolutionarily relevant patterns of information in the environment? Put differently, the question is whether or not organisms may, of their own accord, discern things in the environment that evolution cannot, and whether or not the information so discerned can be subject to an evolutionary analysis. The only possible way in which individuals might be able to discern types of information in the environment that evolution could not is if they had access to a further category of information, which is denied to evolution. We can outline such a category by contrasting it initially with the other two.

Category 3:
These types of situations involve conditions and contingencies that are *not* fixed over the course of time and may change *significantly* from one generation to the next. Each condition and contingency is virtually free to change in any direction during evolutionary time. Moreover, there will be no limits on the range of variation that can keep parameters locked within a certain variance. Because the changes are virtually free and unconstrained, these environmental regularities may not be constant over evolutionary time. We can call these environmental regularities *unpredictable*.[37]

The most central effect unpredictability has for the cognitive structure is that it prevents evolution from developing cognitive modules. There is not enough recurrent predictability in these environmental regularities to furnish evolution with the necessary means for producing specialised solutions. But the challenge unpredictability causes for the survival and reproduction of organisms may nevertheless be crucial. There is one hint that can clearly testify in favour of the importance of unpredictability for the organisms' survival.

Many species possess general learning abilities to learn about the fluctuating regularities of their local environments in order to use the collected information to their best advantage. Thus general learning abilities are the cognitive mechanisms that are best equipped to deal with those aspects of the environment that are

37 This is not to say that these environmental regularities are not predictable within a given period and locality. They clearly must be if organisms are allowed to extract information from them by learning. Unpredictability in this context can only mean that the environmental regularities may change significantly from one generation to the next, from one period to the next, or from one locality to the next. Since evolution is a process that works over long periods of time, it will be blind to these sorts of unpredictable environmental regularities.

marked by unpredictability and novelty. The constancy with which these general learning abilities develop in so many species tells something about the significance of unpredictability for the organisms' survival and reproduction. Even though unpredictable environmental regularities could not furnish the basis on which evolution could have built cognitive modules, evolution needed to find other solutions to learn about unpredictability: evolution created general learning abilities. Now the preoccupation with unpredictability imposed characteristic constraints on the nature of these general learning abilities. The most prominent among those constraints is that general learning abilities could not use content-dependent principles. This is because, firstly, the domains of general learning abilities are not conducive to the development of content-dependent principles, and, secondly, such content-dependent principles would only disrupt the proper functioning of general learning abilities whose essential function is to find application in a wide range of different domains. As it turns out, general learning abilities require a kind of *openness* to deal effectively with unpredictability; an openness that content-dependent principles *by their definition* could not secure. This openness can only be guaranteed once content-*independent* principles are accorded to general learning abilities. Thus it is by virtue of their content-independent nature that these learning abilities can be used across a wide range of different domains. General learning abilities thus clearly belong to general-purpose cognitive mechanisms.

So we can claim that whenever organisms are confronted with unpredictable information that is relevant to them, and in addition cannot apply cognitive modules to deal with it, they must resort to whatever content-independent general learning abilities they possess. The most common examples of such general learning abilities are trial-and-error and instrumental learning.

With the introduction of these general learning abilities we have not yet reached the explanatory target of the evolutionary analysis. The consideration above applies as well to general-cognitive mechanisms of type 1 to which the general learning abilities such as trial-and-error and instrumental learning belong. We have already made clear in section 3.1 that general-purpose cognitive mechanisms of type 1 are not the explanatory target of an evolutionary analysis because everyone agrees that such mechanisms are part of the mind's cognitive structure. The real controversy is whether there are general-purpose cognitive mechanisms of type 2. The consideration above, however, identified unpredictability in the environment as the proper type of information to which general-purpose cognitive mechanisms generally respond.

This alone does not mean that general learning abilities (general-purpose cognitive mechanisms of type 1) do not play a role in an evolutionary account of general-purpose cognitive mechanisms of type 2. They could play such a role if the evolution of general-purpose cognitive mechanisms of type 2 somehow depended on the proper functioning of the general learning abilities. In the following section, I spell out such a story in detail.

5. Evolution of general-purpose cognitive mechanisms

The last section acquainted us with the fact that *unpredictability* in the environment creates relevant challenges for organisms. How organisms deal with unpredictability has consequences on whether they succeed in enhancing their reproductive success. The general learning abilities we have named so far, namely trial-and-error and instrumental learning, seem to use content-independent principles to solve the challenges. Because of their content-independent nature, however, the behaviour the general learning abilities are able to produce is less complex and sophisticated when compared to the behaviour caused by cognitive modules. We are already familiar with this reasoning from chapter I, where we saw how evolutionary psychologists use complexity and sophistication in behaviour as a sign for the presence of cognitive modules and their content-dependent principles.

Now complexity or sophistication in behaviour is not always easy to characterise. Some scientists suggest that we should look at the size of the action repertoire of an organism or the mean number of action elements used for any given task to estimate the complexity of animal behaviour (Byrne, R.W., Corp, N. and Byrne, J.M.E., 2001). But the most illuminating way to estimate the complexity or sophistication of behaviour is to focus attention on the hierarchical organisation of *action-programmes.* Action-programmes typically accomplish their main goal by hierarchically organising lower-level actions that realize their own sub-goals so that each lower-level action contributes its share in the accomplishment of the main goal. Thus a given behaviour will be more complex than another one if it entails a greater number of lower-level actions that are organised hierarchically in a particular action-programme. Richard Byrne (1995) suggests that, for instance, gorillas' manipulations of thistles shows high manual dexterity (coordination of two hands and among individual digits) and hierarchical organisation typical of action programmes. Steven Mithen (1996) believes that the production of stone tools in early humans depends on the hierarchical organisation of the various steps involved in the handling and chipping of stones. Both scientists argue that it is unlikely that these complex action-programmes are uniquely due to general learning abilities such as trial-and-error or instrumental learning.

However, the weakness of general learning abilities with regard to the production of complex and sophisticated behaviour may be their strength once we consider them in the contexts of modular minds. To appreciate their advantage, we need to reconsider some basic facts about modular minds. A modular mind, as we already know, is composed of multiple modules working in relative isolation from each other due to their informational encapsulation. Each module is activated by its specific type of information and reacts to it by a specific complex behaviour. But what happens if problems arise at the modular interface? Before we can answer this question, we need a grasp on how to characterise these problems at the modular interface.

Situations and the problems embodied in these situations belong to a *modular interface* if the solution to those problems would benefit from the combination of the specialised principles trapped within different cognitive modules,[38] but which the mind cannot freely access and combine because of the modules' informational encapsulation (Mithen, 1996). The problem is significant enough for the organism's survival to put considerable pressure on the organism's cognitive resources to find an appropriate solution. But because these cognitive resources cannot easily combine the specialised principles that are trapped within modules, it is typically the general learning abilities that have to take over and find a solution to them. Moreover, since the general learning abilities are responsible for the organism's behaviour at the modular interface, the behaviour appears to be far simpler than the behaviour typically caused by cognitive modules because of the considerations for their content-independent nature. But the simpler behaviour due to general learning abilities is nevertheless necessary to sustain and support the mind at its various modular interfaces lest it break down when confronted by these types of problems. General learning abilities thus fulfil a necessary economic function for the whole structure of the mind.

38 Just because a solution to these problems *would* benefit from the combination of the specialised principles in cognitive modules does not mean that such combination will *actually* occur. On analogy, just because it *would* benefit human beings to fly does not mean that evolution will *actually* produce flying human beings. Given the constitution of human beings, that could be impossible. What I want to say instead is merely that problems at the modular interface create some tension towards combining specialised principles in their solution. If evolution will *actually* produce such combination and under what conditions it could do is the open question I want to investigate. I am specifically interested in the kinds of conditions that need to be satisfied if such combination has to be plausible. Later on I will discuss a concrete example of a modular interface problem also discussed by Mithen.

We are now ready to link the two observations we have made about general learning abilities. Earlier we said that the adaptive function of general learning abilities is to deal with situations involving unpredictable information. Now we found that general learning abilities are necessary to sustain the functioning of the mind at its modular interface. Moreover, we can question whether these two functions of general learning abilities are somehow correlated with each other. Thus it could be that the two functions of general learning abilities stand to each other as *adaptation* and *exaptation*. In one possible scenario, general learning abilities' could have evolved initially to deal with unpredictability, and only later would have taken on the function of sustaining the mind at its modular interfaces. However, I do not believe that one function could have suppressed the other completely. Even if sustaining the mind at its modular interfaces is an exaptation of general learning abilities nowadays, its initial function of dealing with unpredictability must still be in place. So these two functions cannot be completely independent of each other.

Putting these observations together we can propose the tentative hypothesis that unpredictability most likely affects the mind at its modular interfaces. In the context of a modular mind, information of category 3 will most likely be found at the modular interface where the mind cannot use its specialised principles to deal with unpredictability. Moreover, situations at the modular interface are situations that put some pressure on the mind to combine the knowledge hidden away in its different modules to solve these situations in the most beneficial ways for the organism. A concrete example will help to clarify this last point.

Steven Mithen's (1996) analysis of hunting-tools of early human beings may be seen as a discussion of an example of what I have called an interface problem and may illustrate what I have in mind above. Mithen emphasises in his discussion a puzzle about the nature of tools of early humans. The puzzle consists in the fact that early humans did not manufacture tools designed for specific purposes. Analyses of typical tools of early humans have shown that these tools were basically used for general purposes in a wide range of tasks. Why were they not able to manufacture special-purpose tools instead? As Mithen notes, the general-purpose nature of these tools is conspicuously manifest in spear points. Spear points used by early humans show hardly any variability in size or form even though early humans hunted many different types of game. This uniformity contrasts strongly with the great diversity one sees in spear points used by later modern humans – which suggests that at least these later humans were capable of manufacturing specialised hunting tools for particular types of game. Why were early humans not capable of inventing special-purpose tools such as special-purpose spear points?

Mithen's (1996) solution to the puzzle is easy and very elegant. Early humans did not manufacture specialised tools for hunting specific game because their sophisticated knowledge of tool-making and of animal behaviour and anatomy were isolated from each other due to their inclusion in different cognitive modules. Early humans possessed a technology module that was responsible for their complex behaviour in constructing stone tools. They needed to have a specific sort of cognition that would allow them to represent the final form of the tool while they were engaged in organising the many sub-actions needed to produce the tool into hierarchically organised programmes (Byrne, R.W., Corp, N. and Byrne, J.M.E., 2001; Mithen, 1996). But they also possessed a biology module that was responsible for their sophisticated knowledge about and effective exploitation of their natural habitat (Atran, 1990). However, those two cognitive modules were isolated from each other due to informational encapsulation. Confronted by the need to create hunting-tools drawing on the specialised knowledge of both modules, early humans foundered. What they did instead was resorting to their general learning abilities to come up with devices they could use when hunting. Early humans' reliance on general learning abilities also explains the simplicity and uniformity observed in spear points in contrast with the sophisticated stone artefacts that early humans were capable of manufacturing at the same time in other domains.

It is easy to see how the construction of specialised hunting tools informed by biological knowledge would have benefited early human beings in their survival and reproduction. Had they been able to combine their specialised knowledge about tool making and animal behaviour and anatomy, they would have been much better hunters – since they would have created very effective spear points. As a matter of fact, however, early humans could not do this because their cognitive architecture did not allow specialised knowledge to combine freely. But the selection pressures were nonetheless putting considerable strain on the mind to combine the modules' knowledge in producing specialised hunting-tools. An indication of such selection pressures is the fact that later human species developed the required abilities to combine the specialised knowledge to enhance their hunting-tools.

This much is or should be intuitively convincing. Problems at the modular interface force organisms to adjust their cognitive resources to find solutions to these problems as these problems have direct effects on the organisms' survival. Moreover, it appears that solutions that could partly combine specialised knowledge could give organisms a more effective way of solving these problems. So pressure ensues at the modular interface that eventually leads the mind to solutions that combine modules more effectively in an attempt to find advantageous responses.

I need to argue how, on the basis of such pressures, general-purpose cognitive mechanisms of type 2 which profit from content-dependent principles might have evolved, and what the facilitating conditions for such evolution might have been. I will use Mithen's account as a foil in developing my own account. In the following I shall thus briefly outline Mithen's (1996) model.

5.1 Mithen's model

Mithen (1996) believes that there are three phases in the evolution of the human mind:

Phase 1 contains minds that possess only general intelligence such as trial-and-error and instrumental learning, which is used across a wide range of different domains. However, general intelligence, Mithen says, is capable of producing only relatively simple behaviour because of its content-independent nature.

Phase 2 contains minds whose reliance on general intelligence is supplemented by different "special intelligences", each of which is devoted to a specific cognitive domain, and each of which works in isolation from all others. Because these latter cognitive capacities profit from the advantages of content-dependent principles, their learning is much faster and their performance less subject to error in comparison to general intelligence. It is due to the workings of these cognitive capacities that organisms first demonstrate sophisticated behaviour in specific domains. Mithen assumes that there are likely to have been four special intelligences crucial to our evolution:

1. Social intelligence concerned with social cognition, and used for interaction with other con-specifics.

2. Natural intelligence concerned with natural history, and used for understanding of the surrounding natural world, its flora and fauna.

3. Physical intelligence concerned with physical cognition, and used for manufacturing and manipulation of stone and wooden artefacts.

4. Linguistic intelligence concerned with linguistic cognition, and used for communication among other con-specifics.

A critical design feature of the special intelligences is that they work in virtual isolation from each other – they are a form of cognitive modules after all. Knowledge pertaining to different cognitive domains cannot be combined since the special intelligences are closed off from each other. What the cognitive architecture of phase 2 shows is patches of sophisticated behavioural performance due to special intelligences in a wasteland of simple behaviour due to general intelligence.

But general intelligence still retains an important and vital position within the cognitive economy of the mind of phase 2. This is especially pertinent when considering problems that arise at the domain-interface of different special intelligences. When a single solution is required to problems located between special intelligences, and when the search for such a solution would benefit from the knowledge encapsulated in the different special intelligences, the mind must rely purely on general intelligence since the special intelligences are closed off from each other. These last observations give credence to the modular interface problems discussed above.

Phase 3 consists of minds that display an important architectural novelty: Mithen aptly calls it *cognitive fluidity*. Now cognitive fluidity means that knowledge once trapped in special intelligences can be freed up to combine with other knowledge stored elsewhere in the mind. One can understand how, under cognitive fluidity, completely novel ways of thinking and behaving emerge that are dependent on the combination of cognitive modules. However, the cognitive mechanisms enacting cognitive fluidity cannot themselves possess a special-purpose nature since their proper concern is with accessing and combining knowledge that is encapsulated in cognitive modules. If those cognitive mechanisms contained special-purpose principles, their capacity to freely access and combine cognitive modules in whatever ways prove successful in solving particular problems would be strongly jeopardized. This is because the main function of special-purpose principles is to tightly constrain the type of information that can be accessed and reasoned about; this is precisely what those cognitive mechanisms cannot allow.

The conclusion therefore is that those cognitive mechanisms that are responsible for cognitive fluidity are general-purpose cognitive mechanisms. Moreover, as they can profit from content-dependent knowledge because they have access to cognitive modules containing it, those cognitive mechanisms belong to general-purpose cognitive mechanisms of type 2 and are the explanatory target of the evolutionary analysis.

The consideration above brings out the proper explanatory target of the evolutionary analysis, but they do not yet explain how the general-purpose cognitive mechanisms could evolve; how cognitive fluidity first originated in the mind.

Mithen thinks that the answer to the question lies within changes in the nature of language. He adapts a thesis from Robin Dunbar's (1993) to support his arguments. According to Dunbar, language initially evolved as a social tool for exchanging social information in large and complex groups. It most probably first evolved to supplement grooming, but in time ended up replacing it more and more. Originally, then, language evolved as a special-purpose tool for social interaction, and only much later did it transform itself into the general-purpose tool we all know and are familiar with today in which all kinds of information find expression.

But even at the time when language was predominantly about social information, in accord with its evolutionary origin, Mithen suggests that there would nevertheless have occurred *snippets of language about the non-social world* such as tool making, animal be–haveiour, foraging, etc. Mithen further speculates that these *snippets* of language about the non-social world would have had two origins:

Firstly, the snippets about the non-social world could have originated from products of general learning abilities. We know from the consideration above that general learning abilities take control whenever problems arise at the modular interface and cognitive modules cannot solve them. Since the snippets do not belong to the proper domain of the language module, it must have been the general learning abilities that made it possible for early humans to associate some of their vocalisations with specific non-social events (tool handling, animal behaviour, etc.) – and that would have thereby furnished language with foreign elements. The snippets about the non-social world would have been few in number and lacking much grammatical complexity (due to their origin), and they would have strongly contrasted with the overwhelming majority of grammatically complex utterances about the social world (Mithen, 1996).

Secondly, the snippets about the non-social world could have originated from by-products of the cognitive architecture itself. For even though cognitive modules are defined as informationally encapsulated from each other, the degree of encapsulation may not have been capable of completely suppressing interactions between them. Some thoughts or procedures proper to one cognitive module may have entered other modules. These foreign items, however, may not have worked effectively in the context of the invaded module. This is because the foreign items would have incorporated structures unknown to the invaded module. At this stage of evolution, the degree of encapsulation would have prevented the cognitive modules from working together successfully. Only in the case of the language module, it could have been possible that the foreign items would have been available for expression in language.

Whatever the source of the snippets of language about the non-social world, the snippets would have entered receiving minds as parts of the general flow of social language. As such, they would have been decoded by specialised linguistic procedures and interpreted by social intelligence.

Now Mithen argues that those organisms that were better at gaining knowledge about the non-social world by exploiting the relevant snippets of language acquired a considerable selective advantage over other group-members that could not do this. In addition, those organisms that improved the production of the relevant snippets of language or the decoding of them would have increased their benefits even more. In any case, once evolution had started to select the organisms that could better exploit the relevant snippets, a definite trend towards the integration of cognitive modules would have been the outcome. As a consequence of this, language would have quickly changed from a special-purpose device dedicated to social information to a general-purpose device employed in the dissemination of all kinds of information.

To sum up Mithen's proposal we can claim that he sees language as the main promoter of cognitive fluidity because it acts as a medium for the integration of cognitive modules. With cognitive fluidity, general-purpose cognitive mechanisms of type 2 come finally into view.

5.2 A new proposal

I believe that Mithen is correct in his basic ideas about the evolution of general-purpose cognitive capacities of type 2. But there is a major problem, which prevents me from using his approach directly for my purposes. As Mithen's account depends essentially upon the presence of a linguistic ability, his account is directly applicable only to human beings. But I intend to give a much broader account of the evolution of general-purpose cognitive capacities, one that could apply equally to animals that do not posess linguistic abilities. So if I want to use some of Mithen's basic ideas, I will need to rework his approach in significant ways.

My proposal retains the following features from Mithen's proposal: firstly, it retains the emphasis on the considerable selection pressures present at the modular interface, and, secondly, it also retains the vital role general learning abilities qua general-purpose cognitive mechanisms of type 1 play in producing relatively unsophisticated behaviour at the modular interface; this is in comparison to the sophisticated behaviour based on complex action programmes due to cognitive modules. But my proposal cannot, as indicated, retain the essential role that language plays in bringing about general-purpose cognitive capacities of type 2. So it must find a substitute for the work language does in Mithen's proposal.

As a first introduction to this substitute, I want to bring out the essential ideas of Mithen's proposal more clearly. Note that Mithen assumes that it is grammatically *simple* snippets of conversation about the non-social world that, whilst few in number, drive selection towards general-purpose cognitive mechanisms. Even though the production mechanisms of those snippets of conversation, i.e. the general learning abilities, are different from the normal production mechanisms of social language, i.e. the language module, these snippets are still able to invade and embed themselves in social language. Once embedded in social language, these snippets of conversation profit from the linguistic operations that decode them and the social interpretations that interpret them by hijacking social language's normal transmitting mechanisms. As a consequence of this, they enter the receiving minds of other individuals as part of the general flow of social language. So Mithen's proposal depends on three basic conditions:

(1) *Production Mechanisms*:
Relatively unsophisticated target behaviour, i.e. snippets of conversation about the non-social world, is produced by general learning abilities. In addition, this target behaviour is able to invade the sophisticated behaviour caused by a cognitive module, i.e. the social language with its sophisticated grammatical structure.

(2) *Transmission Mechanisms*:
The target behaviour, i.e. snippets of conversation about the non-social world, seizes the normal transmission channels of the relevant cognitive module from sender to receiver, i.e. linguistic decoding and social interpretation.

(3) *Selection Assumption*:
The target behaviour is appropriate to produce selection for general-purpose cognitive mechanisms that facilitate the integration of cognitive modules.

As I said, my proposal takes on Mithen's basic ideas, but introduces different conditions to account for non-linguistic organisms. So I shall now outline different production and transmission mechanisms. Let me briefly introduce the ideas before I describe them in more detail.

According to my proposal, cognitive modules are sometimes inappropriately activated in specific circumstances by information at the modular interface that does not belong to the cognitive module's proper domain. When this happens, the cognitive module cannot process the information efficiently and this may lead to sub-optimal behaviour that is markedly different from its typical output – given that the module is capable of producing an output at all under those circumstances. But in case the cognitive module produces the sub-optimal behaviour, the behaviour will reflect the partial influence of the module's content-dependent principles. This means that the target behaviour will manifest certain degrees of complexity and sophistication, e.g. some forms of hierarchical organisation typical of action-programmes that is not generally attainable on the basis of general learning abilities alone.[39] In the light of all this, the new behaviour may confer significant advantages on the organism producing it and as a consequence of this the target behaviour would draw the attention and interest of other group-members observing the benefits. However, because the target behaviour is produced by an *inappropriately* activated module, the transmission of the target behaviour and its benefits from sender to receiver in the group cannot count on the cognitive module's normal transmitting channels. The only remaining cognitive capacities capable of dealing with the target behaviour, i.e. exploring the behaviour, learning about it, and emulating it in the group, are the

39 The higher complexity of this behaviour is, as we said, due to the partial influence of content-dependent principles; these principles, in their turn, are the "sediments" of statistical information that is not accessible by general learning abilities (see evolutionary psychologists' argument from statistical information in chapter I). The behaviour will therefore, as far as it shows some influence of content-dependent principles, stand out from the rest of the behaviour produced by general learning abilities because it will include elements that could never have been brought about by these general learning abilities alone since the latter abilities do not have access to the relevant statistical information.

general learning abilities. I suggest that the target behaviour can be transmitted in the group by some form of social learning.

My proposal in some ways stands Mithen's proposal on its head. While Mithen claims that it is relatively *simple* snippets of conversation about the non-social world that invade the relatively *complex* utterances directly produced by the language module, my proposal claims instead that it is relatively *sophisticated* target behaviour that invades the relatively *simple* behaviour directly produced by general learning abilities. So my proposal depends on the following conditions:

(1) *Production Mechanisms*:
Relatively sophisticated target behaviour is caused by the inappropriate activation of modules at the modular interface. This target behaviour is able to invade the behaviour caused directly by general learning abilities.

(2) *Transmission Mechanisms*:
The target behaviour seizes upon some forms of social learning to spread from sender to receiver.

(3) *Selection Assumption*:
The target behaviour is appropriate to produce selection for general-purpose cognitive mechanisms that facilitate the integration of cognitive modules.

The mechanisms responsible for producing the target behaviour and the mechanisms responsible for transmitting it in the group need further clarification to substantiate the proposal. I start first with the production mechanisms.

As was asserted above, cognitive modules can sometimes be activated in inappropriate ways. A concrete example of such an inappropriate activation will further elucidate this important point. Let us imagine an organism, actually an early human being that possesses, among other things, one module for technical skills and another module for handling biological information about animals; their foraging behaviour and anatomy (Atran, 1990; Mithen, 1996). The modules generally work in relative isolation from each other, but in some special and rare circumstances, animal parts such as bones and antlers (which are normally handled by the biology module) may have activated the module for technical skills. In this context, Mithen (1996) notes that in the rare instances in which early humans did work animal bone, they handled it as if it was stone, thereby engaging in an inappropriate method of working bone. That is precisely what one should expect.

Activated by input that does not belong to the proper domain of the technology module, the module cannot work efficiently because it will try to apply its specialised principles and operations on material that is not suitable for such handling. Even though chipping may not be the most efficient method to work bones, it most probably would have given rise to useful and effective tools due to the physical properties of bones that are capable of sustaining great impacts without fracturing. Thus bones would have manifested the influence of content-dependent principles characteristic of the technology module: they would have shown signs of complex action-programmes that implicated both memory systems capable of retaining templates of the finished tool and exceptional manual dexterity involved in the coordination of the two hands and among individual digits.

On any account, the bone chipping would have introduced a genuinely novel behaviour that could not have been caused by the general learning abilities in the absence of the inappropriately activated technology module. What is more, the behaviour of bone chipping may have constituted a first instance of the partial integration of the technology module with the biology module. This partial integration, however, could not have been considered stable, at least not at the time of the "invention" of bone chipping. For we must take into consideration that the cognitive architectures of most group members who noticed the bone chipping would have exerted a strong resistance against the integration of cognitive modules. First confronted by bone chipping, group-members would initially have been at a loss as to how to interpret or copy this peculiar novel behaviour. The same kind of consideration may even have applied to the original inventor of bone chipping at a later time when he needed to repeat his invention. This is because the inventor's own cognitive architecture would similarly have exerted a strong resistance against integration. The next time he would have wanted to repeat the invention, he may not have "remembered" how to perform the necessary steps.

Since tools made of bone possess excellent structural properties, we can presume that organisms that were capable of producing and using such tools would have benefited greatly compared to other group-members that could not. There would have been an incentive in the group to retain this kind of behaviour and even to instantiate it more systematically. But because the specific behaviour would have been produced by an inappropriate activation of a cognitive module, it might not have been possible for the behaviour to be transmitted by the typical channels of the module. The only remaining cognitive mechanisms that could have been used in transmitting the target behaviour would have been the general learning abilities, i.e. trial-and-error or instrumental learning on the basis of reinforcement and punishment. Under these circumstances, we can imagine group members observing the positive effects that the new tool has on the inventor, which they then try to copy in whatever ways possible to them.[40] In their attempts at copying the behaviour they would have been guided by their general learning abilities. So although the target behaviour is induced externally from general

40 These considerations assume a general urge and ability in animals to try to copy. This assumption seems to be unproblematic for animals living in social groups. Social animals can profit at low cost from findings and inventions of other group members if they have learning abilities. Learning a successful strategy from others is likely to be cheaper in energy expenditure than going about to invent a new strategy for oneself. The claim that such general urge to copy is also to be found in non-social animals is more contentious, but does not greatly affect my consideration here.

learning, it will have to use general intelligence as a vehicle of transmission to spread in the group.[41]

Moreover, as is the case with simple snippets of conversation about non-social information in Mithen's model, those organisms that could systematically exploit the target behaviour to increase their knowledge of new ways to interact with their environment would have experienced considerable selective advantage in relation to other group-members that could not do so. In time, selection could have brought about a continually more integrated cognitive architecture that could combine various cognitive modules to issue in more sophisticated behaviour: the end product of such a selection could then have led to general-purpose cognitive mechanisms of type 2.

To summarise, my proposal suggests that the following conditions need to be in place to lead to the evolution of general-purpose cognitive mechanisms of type 2:

(1) General learning abilities typically control the behaviour at the modular interface and give rise to relatively simple behaviour when contrasted with behaviour caused by cognitive modules.

41 Incidentally, the transmission of relatively complex behaviour by general intelligence might also explain one of Mithen's (1996) puzzles over technical conservatism in early humans. Namely, it explains why early humans ignored bone, antler and ivory as raw materials. Given that relatively complex behaviour within the sphere of influence of general learning abilities cannot resort to modular transmission channels, early humans would have had to use the much weaker transmission channels of general intelligence. As indicated earlier, the operations of general intelligence can only give rise to behaviour that is slow and error-prone (i.e. only simple behaviour is possible). So it might have been the case that, over and over again, advantageous behaviour would have been unfaithfully copied or lost altogether. So because of the poorer transmission quality of general intelligence, novel behaviour due to the inappropriate use of cognitive modules might have had a hard time establishing itself in the population. This fact could similarly explain the fact that bone tools do not exist in the archaeological record.

(2) Inappropriately activated cognitive modules sometimes cause atypical behaviour that manifests the partial influence of the module's content-dependent principles. Because such relatively sophisticated behaviour is caused by an inappropriate use of the module, the behaviour cannot count on the typical transmitting channels of the module.

(3) Although the general learning abilities do not cause the target behaviour, the behaviour is nonetheless evaluated and accepted in the behavioural repertoires by these general learning abilities. Organisms have to learn about the value of the novel behaviour by closely monitoring the consequences such behaviour has by using their general learning abilities to emulate it.

(4) Those organisms that could exploit the novel target behaviour more efficiently would gain a considerable selective advantage over other group-members that could not do it. Since some target behaviour constitutes a partial integration of cognitive modules, the demand on the efficient exploitation of it leads to improvements in the interaction among cognitive modules and thereby facilitates the development of general-purpose cognitive mechanisms of type 2.

But *how* can cognitive modules be inappropriately activated in order to give rise to the target behaviour? In the following I outline possible ways of activating cognitive modules in this specific way.

(A) Mechanisms producing the target behaviour

The clue in understanding how cognitive modules can be tricked into inappropriate activation lies in how cognitive modules work. A reminder of Sperber's views on modularity will illuminate this point. We saw in chapter I that Sperber (1994) claims that the modules' input condition or domain is determined by the internal structure of the cognitive module itself and its relations to other modules. Recall that Sperber introduces the concepts of a *proper domain* and an *actual domain* to analyse these input conditions.

The *proper domain* of a module is all the information in the organism's environment that satisfies the input conditions and is the module's biological function to process.

The *actual domain* of a module is all the information in the organism's environment that may satisfy the module's input conditions, whether or not it also belongs to the module's proper domain.

In addition to the two domains mentioned in chapter I, Sperber also introduces a *cultural domain* of a module. The *cultural domain* of a module is all the information in the organism's environment that has been *purposefully* fashioned to satisfy the module's input conditions. The cultural domain of a module is a sub-category of the actual domain, which extends the module's range of application even further than its proper domain. Humans, for instance, presumably have a special module that is used in explaining and predicting actions of real persons by imputing mental states to them (Baron-Cohen, 1995; Leslie, 1994, 1995). Humans also have the ability to purposefully change their environments to fit the input conditions of specific modules. In fact, humans have done so by inventing television, movies, and films the perception of which equally activates our theory of mind modules: we treat figures on the television screen as if they were persons and ascribe mental states and motives to them.

These different domains are relevant for my purposes here in the following ways. We suggested that the mind experiences considerable selection pressures at its modular interface. Organisms will have to do whatever is in their cognitive power to alleviate the accruing strains by finding strategies to reduce the strains. One strategy, and indeed the one used more often by organisms, is to use general learning abilities such as trial-and-error or instrumental learning to produce adaptive responses that solve these modular interface problems. But we can speculate that sometimes the pressure on the cognitive architecture to find a suitable response to the modular interaction problem gets so strong that the mind reacts with an inappropriate activation of a cognitive module in an improper domain. We saw how in early humans, for instance, cognitive modules for technical skills might have been used on some occasions for manipulating animal parts that constituted the proper domain of a biology module. It is true that modules may not work effectively when this happens,[42] but in the cases in which behaviour occurs, e.g. bone chipping, it will be more sophisticated than behaviour produced by general intelligence alone. Such inappropriate uses of cognitive modules can give rise to *inventions* that introduce genuinely novel patterns of behaviour not previously observed in the population. Here it is that the different domains of cognitive modules play a role in bringing about inappropriate uses of cognitive modules.

Inappropriate Uses of Cognitive Modules by Actual Domains

A cognitive module may be inappropriately activated by information of the actual domain. As we saw above, a cognitive module's actual domain, indeed also its cultural domain, is constituted by information that was not selectively intended for processing by the module – this is because the information does not belong to the proper domain of the module. But even though this information was not originally intended for processing by the module, it does nevertheless satisfy the module's input conditions, and as a consequence tricks the module into action. We can further assume that such information may originally belong to a *different* cognitive module, as is the case in Mithen's example of the chipping of bones. There we saw that animal bones belong to the biology module, but that under special circumstances specific properties of bones were apparently capable of activating the technology module. The upshot of these observations is that similar kinds of situations could occur in different domains and would lead there to inappropriate activations of cognitive modules. As a consequence, these inappropriate activations of cognitive modules cause a constant influx of

42 But even if they do not work effectively, some of their produced behaviour may still have beneficial effects on the organism.

novel and relatively sophisticated behaviour that could no longer count on the regular modular processes for the interpretation of it; this is because the behaviour results precisely from inappropriate activations of cognitive modules. Confronted with such novel behaviour, organisms will have a hard time to understand or learn it.

This latter claim, however, depends on how common the actual information triggering the cognitive module is. For if the actual information is very common in the environment of organisms possessing the relevant module – as is the case, for instance, with television and the theory of mind module in humans – no special problems arise for understanding or learning this novel behaviour. The common actual information will activate the cognitive module almost automatically. But if the actual information is sparse and infrequent, and if, moreover, a certain practice is needed to sensitise oneself to the impact of it, problems may arise in understanding and learning about the novel behaviour precisely because the usual modular procedures are not available. Let us suppose that a single organism detects a new and rare food such as nuts that by chance satisfy the input conditions of its foraging (or taste) module, but which also requires particular technical handling. Other individuals perceiving the individual opening the nut could not automatically resort to the required handling. They would have to resort instead to their general learning abilities to understand and learn the novel behaviour.

Inappropriate Uses of Cognitive Modules by Trade-Offs

Another way to bring about inappropriate uses of modules results from the trade-off decisions between speed and efficiency embodied in the designs of modules. Cognitive modules are an optimal solution wherever cognitive capacities are needed that cannot be allowed to engage in time consuming and unreliable computations. For the amount of time an organism spends on a given task is an important constraint on the resources of the organism as a whole. Thus cognitive modules can be fruitfully viewed as trade-offs between speed and efficiency. As a consequence of such trade-offs, they will be prone to false positives: being triggered while the appropriate stimulus is not present. Under resource limitations, it pays the cognitive module more to be safe than sorry. A similar consideration about commonness or scarcity of the relevant information in the environment applies here as it did in the previous observation of the actual domains of modules. It is clear that the more common a piece of information is in the environment, the greater the chances that a cognitive module overshoots.

Moreover, since cognitive modules incorporate trade-offs between important parameters, the ability of misfiring is a highly context-sensitive matter and might not occur with the same frequency in different domains. The misfiring of a cognitive module does not always have to lead to negative influences on the organism. On some occasions, by mere chance, it may lead to genuinely novel behaviour that improves the organism's way of interacting with its environment. If other group-members were drawn to this kind of behaviour because of its advantages, they would have to resort to general learning abilities to understand and learn it since it might be difficult for them to recreate the same misfiring of the cognitive module. As an aside, it might even be that bone chipping in early humans that Mithen (1996) discusses was ultimately due to the misfiring of the technology module in the context of bone material.

Inappropriate Uses of Cognitive Modules by Experience

A final way to bring about inappropriate activation of cognitive modules is with the important role inferences play in the triggering algorithms of cognitive modules. Let me explain how this works. Each cognitive module possesses a characteristic triggering algorithm that screens the incoming information for a set of cues that are correlated with the presence of some relevant situation to which the cognitive module applies (Gigerenzer, 1997). Now some cognitive modules possess very simple triggering algorithms that screen for only a small set of perceptual cues, e.g. face recognition modules whose triggering algorithms check the incoming information for specific facial configurations (Slater and Quinn, 2001). But some other cognitive modules possess more sophisticated triggering algorithms that need to attend to more complex sets of cues to infer the presence of relevant situations, e.g. cognitive modules that detect the status of a good mate by triggering algorithms that screen the incoming information for such diverse cues as reputation, health, propensity to commitment, and generosity (Buss, 1999).

While the simpler modules are activated almost automatically on detection of the perceptual cues, it seems likely that in the more complex cases something like experience is needed to draw the inference from the cues to the relevant situation. In the context of inappropriate activations of cognitive modules, it is important that the dependence on inference leads the organism sometimes to believe that the relevant situation is present when it in fact is not. What actually triggers the module in such a case is not so much the real presence of the situation in the environment as the situation's *imagined* or *perceived* presence. Some of the inappropriate activations of the cognitive module, however, may lead to beneficial behaviour for the organism. As in the other cases, group-members that are drawn to the novel behaviour because of its benefits need to apply their general learning abilities to interpret and learn this behaviour for themselves since the context may not make it easy for them to draw the same false inference as the other organism.

In summary, these three ways of bringing about inappropriate activation of cognitive modules constitute the production-mechanisms necessary for the causation of the relevant target behaviour. But the production-mechanisms of the target behaviour are only the first part of the analysis of the evolution of general-purpose cognitive mechanisms. The second part of the analysis is concerned with the transmission-mechanisms that allow other group-members to interpret and learn the target behaviour for themselves. We need to analyse these transmission-mechanisms next.

(B) Mechanisms transmitting the target behaviour

The production-mechanisms are responsible for generating behaviour that reflects some partial interaction between cognitive modules (such as bone chipping). But if this sort of behaviour has to function as the facilitator of the evolution of general-purpose cognitive mechanisms, the target behaviour needs mechanisms with which it can spread in the group where it can act as a selective agent that favours those organisms that are more successful in exploiting it. The most efficient way evolution has to promote the transmission of the target behaviour in the group is by giving group-members an incentive to learn about it. Now organisms in a group are generally attracted towards behaviour that has beneficial outcomes for the agent and they will try to copy the behaviour in order to reap the same benefits for themselves. These organisms will try everything in their cognitive powers to incorporate the observed behaviour into their own behavioural repertoire. If the organisms succeed, they will benefit from its outcomes. If they do not succeed, the behaviour may be lost for them. In any case, those organisms that succeed in incorporating the behaviour in their repertoires will enjoy a selective advantage over others that cannot do this.

We can assume in addition that organisms would enhance their selective advantage even more if they could develop more reliable ways of learning about novel types of behaviour in the group or if they could modify their cognitive architecture in ways that made it easier to develop novel types of behaviour for themselves. The outcome of these improvements of the cognitive capacities of organisms would have ultimately led to growing interactions among cognitive modules resulting in sophisticated behaviour at the modular interface – thus possibly leading in the end to the evolution of general-purpose cognitive mechanisms of type 2.

As this consideration suggests, the evolution of general-purpose cognitive mechanisms of type 2 depends on social learning as the transmitting-mechanisms for target behaviour. I cannot go into the details of the vast literature on social learning in nonhuman animals,[43] but I shall outline different forms of social learning to analyse their consequences for the evolution of general-purpose cognitive mechanisms. Typically, psychologists distinguish three forms of learning in social groups:

Stimulus Enhancement

Learning by stimulus enhancement occurs where an observer, the learner, has its attention brought to salient stimuli by another organism, the model. The model does not have to actively interact with the learner. It suffices that the model performs actions that have positive outcomes. Because of these outcomes, the place where the model is situated and the objects with which the model interacts are automatically made more salient for the learner. In addition, learning by stimulus enhancement does not imply that the learner learns the model's manipulations. On the contrary, the learner in stimulus enhancement, once its attention is brought to salient stimuli, devises its own behavioural responses (which includes whatever behaviour it already has in its repertoire or behaviour it can develop by trial-and-error).[44]

Emulation

Learning by emulation occurs where an observer learns something about the goals to pursue from a model, but does not learn anything about the specific methods the model uses to achieve the goal. For example, an organism may learn to use a tool by observing the effects the tool has for a model, but it does not learn the specific manipulations the model uses to achieve those effects. The learner

43 For a discussion of social learning in animals see Tomasello and Call (1997), and Russon, Bard, and Taylor Parker (1996).

44 An efficient way of exploiting learning by stimulus enhancement is given to social species. For in social species relatives and allies typically stay close together. Should one individual discover a novel behaviour, the chances are high that the other nearby individuals will perceive the relevant event (Tomasello and Call, 1997).

thus seeks to reproduce the same effects, but does not attend to the specific ways in which others manipulate the tool to bring about the effects. The learner will manipulate the tool with its own idiosyncratic behavioural repertoire until it results in the desired effects.

Imitation

Learning by imitation occurs where an observer learns both something about the goals to pursue from its model and something about the manipulations the model uses to bring about the goals. Thus the observer needs to understand both the means and the ends of the model to be capable of learning by imitation. It is generally agreed that true imitation is the ability to learn some behaviour solely from seeing the behaviour performed by another organism in the group. Only true imitation guarantees that the observer is able to learn new behaviour that is not already contained in its behavioural repertoire.

In summary we can state that social learning comes in different degrees. At the one extreme is stimulus enhancement that does not actually lead directly to the learning of an other's behaviour. In stimulus enhancement an organism devises its own behavioural responses to bring about the same effects, but cannot reproduce a behaviour that is not part of its behavioural repertoire. At the other extreme is true imitation that is clearly able to furnish the organism's behavioural repertoire with genuinely novel behavioural options. It therefore follows that the destiny of novel forms of behaviour in the group depends pretty much on the forms of social learning employed in the group. This is shown by the following consideration.

Imagine a situation in which a particular organism "invents" a new way of behaving in specific circumstances that is the result of some interaction between different cognitive modules. We can further assume that this new type of behaviour benefits the inventor in some ways, e.g. by giving it access to precious food. Other group-members observe the positive effects this new way of behaving has for the inventor, and are thus motivated to learn about it. If, however, the other group-members can "learn" about this behaviour only by means of stimulus enhancement or emulation, they may not be capable of producing the same effective ways to gain access to the food. As a consequence of this, the correct behaviour may be lost and cannot act as a selective agent in the group. On the other hand, if the other group-members were capable of true imitation, it would seem easy for them to pick up on the novel behaviour. The novel behaviour may therefore easily spread within the social group and enhance the reproductive success of those group-members that could exploit it.

As these considerations make clear, evolution of general-purpose cognitive mechanisms of type 2 depends not solely on the types of production-mechanisms causing the target behaviour, but also on the types of transmission-mechanisms that are available in a social group. Thus, evolution of general-purpose cognitive mechanisms of type 2 consists in selection affecting both the production-mechanisms and the transmission-mechanisms; leading to a co-evolution of both.

Incidentally, these considerations also explain why many species of non-human animals have not developed general-purpose cognitive mechanisms of type 2. Contrary to what one sometimes finds in the literature, it is not the absence of language per se in non-human animals that accounts for their lack of general-purpose cognitive mechanisms of type 2. The reasons for their lack of such general-purpose cognitive mechanisms might go deeper than that and begin with the nature of social learning skills in group-living animals. There is still an on-going debate in animal psychology, for instance, whether primates demonstrate clear and conclusive signs of true imitation in the laboratory (Tomasello and Call, 1997). The outcome of this debate may have serious consequences for the question about the cognitive structure of non-human animals. For the absence of true imitation could, as we said, prevent the effective spreading of novel behaviour in the group. This by itself does not mean that the cognitive architecture of non-human animals would not be able to produce new ways of behaving; these could still be caused by the outlined production-mechanisms. However, the "weakness" of the animals' transmission-mechanisms may not allow them to stabilise the novel behaviour in the group where it can act as the necessary selective agent. As a consequence of this, novel behaviour may simply be lost and with it an important vehicle of the evolution of general-purpose cognitive mechanisms.

An additional explanation for why not many species have developed general-purpose cognitive mechanisms of type 2 can take on the more familiar claims and point to the importance of language as the vehicle of the evolution of general-purpose cognitive mechanisms. As Mithen (1996) shows in his proposal, language would definitely facilitate the evolution of a cognitively fluid mind. Once organisms developed a fully-fledged linguistic competence, this might have served as a highly efficient way to transmit information about the natural and social world. With language in place, organisms can learn about new ways of behaving even without needing to observe the behaviour, but simply by being told about it. Language could thus have led to an explosion of new ways of behaving or thinking. And once written records emerged, even the slightest innovation could be recorded for other organisms to use. My aim in this chapter, however, has been to show how evolution of general-purpose cognitive mechanisms of type 2 may come about without the benefits of language. What language does is simply speed up and facilitate the process in incredible ways. But the process is also possible in the absence of language.

To sum up, my proposal suggests that there are three conditions that need to be satisfied in order to lead to evolution of general-purpose cognitive mechanisms of type 2. The first condition is that problems at the modular interface sometimes elicit sophisticated behaviour. It is assumed that some of this behaviour generates enough selection pressure to act as the vehicle of selection for general-purpose cognitive mechanisms performing the integration of cognitive modules. This condition I take from Mithen (1996). The second condition is that there are production-mechanisms capable of causing the relevant target behaviour. I have outlined the kinds of production-mechanisms that can generate the target behaviour. The third and final condition is that there are transmission-mechanisms that allow the target behaviour to spread in the group. I have outlined how this condition depends on the forms of social learning. Given these three conditions, one can design a model of how evolution of general-purpose cognitive mechanisms of type 2 could have been possible.

6. Conclusion

The outcome of this chapter demonstrates that weak versions of evolutionary psychology present a consistent and fruitful research programme into the mind's structure. Not only are weak versions of evolutionary psychology able to accommodate the discovery of cognitive modules, but also able to devise evolutionary accounts of general-purpose cognitive capacities. This constitutes a progress when compared to Samuels' library model of cognition. While Samuels' arguments only showed *that* general-purpose cognitive mechanisms are possible, weak versions of evolutionary psychology can also account for their evolution by outlining the conditions conducive to the development of general-purpose cognitive mechanisms. Additionally, evolutionary psychology presents an alternative with its account of general-purpose cognitive mechanisms to traditional cognitive psychology. The aim of this chapter was to investigate the possibilities of such an evolutionary account of general-purpose cognitive mechanisms.

The first observation that guided us was the fact that any analysis needs to bring out its explanatory target. We thus distinguished different forms of general-purpose cognitive mechanisms and found that the most interesting general-purpose cognitive mechanisms from an evolutionary perspective are the ones that can somehow benefit from content-dependent principles employed in cognitive modules.

The second observation pointed out that evolutionary psychologists have concentrated all their efforts on analysing adaptive problems of the outside environment in order to make their case for the massively modular mind. I then suggested that evolutionary psychologists have neglected all problems due to the modular architecture of the mind itself. Faced with the simultaneous activation of several cognitive modules, the mind needs to find adaptive responses to these situations, which are as important for the organism's reproductive success as solving adaptive problems of the outer environment. To distinguish these two different kinds of problems I introduced the distinction between an outer selective environment and an inner selective environment.

In a next step we looked at one recent proposal from evolutionary psychology that deals with these problems of the inner selective environment. This recent proposal tries to deal with those problems on the basis of special-purpose cognitive tools already familiar from our discussion of evolutionary psychology. I raised two objections to this special-purpose solution to show that it is not satisfying. I then suggested a different approach to these issues that takes general-purpose cognitive mechanisms seriously, and that tries to give of them a positive evolutionary account.

Firstly, we declared that any evolutionary analysis of cognitive mechanisms should focus on the types of information that are most appropriate for the cognitive mechanisms under investigation. We suggested that for general-purpose cognitive mechanisms unpredictability in the environment is the most appropriate information to which these cognitive mechanisms respond.

Secondly, we suggested that unpredictability should most likely affect the mind at its modular interface. Since cognitive modules take precedence in domains where predictable patterns of information emerge during evolutionary time, unpredictable information will occur only in domains where cognitive modules cannot extend their special-purpose reasoning. Such domains constitute the mind's modular interface, and problems arising at the modular interface need to be taken care of by non-modular cognitive abilities such as general-purpose cognitive mechanisms.

However, this alone did not suffice to account for the evolution of the most interesting general-purpose cognitive mechanisms of type 2 since we also found that problems at the modular interface can be taken care of easily by the uncontested general-purpose cognitive mechanisms of type 1 such as trial-and-error or instrumental learning. Additional support needed to demonstrate the importance of general-purpose cognitive mechanisms of type 2.

We finally found this additional support by reworking some insights underlying Mithen's proposal of cognitive fluidity. The basic insight of Mithen's proposal points to situations in which the mind applies cognitive modules in domains that are not specific to them, but belong to different modules altogether. The basic idea is that inappropriately used cognitive modules will produce genuinely novel behaviour that carries the marks of the content-dependent principles of the inappropriately used cognitive module. This novel behaviour, however, once produced cannot count on the cognitive module's normal transmission mechanisms and will have to spread in the group on the basis of general-purpose learning abilities.

We suggested that this inappropriately produced novel behaviour would have given rise to significant outcomes that motivated other group-members to exploit the behaviour to their own advantages. Under these circumstances, selection would have set in to exploit this novel behaviour caused by partial integrations among cognitive modules. This kind of novel behaviour would have acted as a catalyst of the evolution of general-purpose cognitive mechanisms of type 2. As selection would have led organisms to integrate more and more of their cognitive modules to either produce novel behaviour or to learn about novel behaviour observed in the group, cognitive fluidity among cognitive modules would have emerged in some species. We concluded the section with some considerations about why many non-human species seemingly did not develop cognitive fluidity to its fullest potential.

This and similar evolutionary accounts of general-purpose cognitive mechanisms lend support to weak versions of evolutionary psychology that do accord general-purpose cognitive mechanisms a place within the mind's structure. Together with their defence in chapter II, and the development of an evolutionary account of general-purpose cognitive mechanisms in this chapter, weak versions of evolutionary psychology are able to present a consistent alternative to traditional cognitive psychology. Additionally, weak versions of evolutionary psychology have substantial implications for other aspects of the study of cognition. One important implication of evolutionary psychology is its assumption of genuine cognitive capacities in non-human animals. The next chapters will be concerned with this specific assumption, and they will analyse and defend it by engaging with proposals developed in the literature.

IV. Evolutionary Taxonomies of Cognition

The outcome of the last chapters is that weak versions of evolutionary psychology represent a consistent research programme for the study of the mind. The goal of the following chapters will be to critically discuss an assumption of evolutionary psychology's approach to the human mind: namely, the assumption that non-human animals have genuine cognitive capacities.

This assumption is based on the logic of the evolutionary process itself. For evolution does not make leaps: the complex designs that evolution produces are built from pre-existing designs via small and incremental steps that lead from one intermediary form to another. Some of these intermediary forms will be simpler or more rudimentary forms of a given design that approximate it in certain interesting respects. Thus closely related species will share a number of homologous bodily traits that carry out the same or relevantly similar biological functions. All primate species, for instance, possess a grasping hand that they use in their foraging ways. Now it is true that some species surpass others in the use of the hand, and that the human species surpasses them all in terms of the hand's anatomical structure and its possible uses derived on the basis of this structure. However, the primate hands generally carry out the same or relevantly similar biological functions, since they are taxonomically related to each other: hands are there for grasping and manipulating objects.

Cognitive capacities do not form the exception to this evolutionary rule. Closely related species will share a number of homologous cognitive capacities that carry out the same or relevantly similar cognitive functions. As with bodily traits, it is true that some species surpass others in the use of cognition and that the human species surpasses them all in terms of the sophistication, complexity and flexibility of its cognitive capacities. However, closely related species may share many cognitive capacities that carry out the same or relevantly similar cognitive functions. So if evolutionary psych–ologists do engage in evolutionary approaches to the human cognitive capacities, they must also assume that non-human animals have cognitive capacities, which are taxonomically related to the human cognitive apparatus.

The goal of the current chapter is to discuss and develop an evolutionary taxonomy of such animal cognitive capacities in order to clarify further evolutionary psychology's assumption. Now evolutionary taxonomies accomplish two objectives. Firstly, they outline and distinguish the different cognitive options that evolution produced in non-human animals, and, secondly, they bring these cognitive options into systematic order of growing sophistication. The motivation behind evolutionary taxonomies is their ability to create synoptic insights into the evolution of cognition by organising our thinking about cognition in systematic ways. Moreover, evolutionary taxonomies support us in framing the intermediary stages of cognitive evolution that led to the development of the human cognitive capacities, and they thus may further our understanding of the latter's evolutionary history.

The chapter is structured as follows: Section 1. analyses Daniel Dennett's proposal of an evolutionary taxonomy. After introducing Dennett's proposal in detail, I shall raise some difficulties that threaten the consistency of his model. Section 2. deals with these problems of Dennett's model by analysing empirical findings of animal psychology. By the end of the chapter, I will have arrived at a detailed evolutionary taxonomy of animal cognitive capacities.

1. Dennett's Tower of Generate-and-Test

Daniel Dennett is not the only scientist that has suggested an evolutionary taxonomy of cognitive capacities in non-human animals.

Other scientists like the cognitive archaeologist Steven Mithen (1996) and the psychologist Luria Vigotsky (1993)[45] have also introduced such taxonomies in their work. But I am concentrating on Dennett because his evolutionary taxonomy is the more influential one in recent discussions about the topic.

1.1 *Dennett's evolutionary taxonomy*

Dennett (1994, 1995, 1996) introduces his evolutionary taxonomy of cognitive capacities through his model of the Tower of Generate-and-Test. The model consists of four different levels each correlated with a different stage in the evolution of the organisms' abilities to adapt their behaviour to their environments. Dennett chooses to focus in his model on the underlying mechanisms that enable organisms to *generate* and *test* hypotheses about their behaviour. As we move from one level to the next, organisms employ more and more sophisticated tools to generate and test hypotheses pertaining to their behaviour.

Dennett is quite explicit about the oversimplifications, which any broad taxonomy necessarily carries with it. His model "is an outrageously oversimplified structure, but idealization is the price one should often be willing to pay for synoptic insight." (Dennett, 1996, p. 81) One oversimplification is that the model assumes clearly demarcated periods in the evolutionary history of cognitive capacities. Such demarcated periods are almost certainly not to be

45 Vygotsky (1993) suggests an evolutionary development of four stages leading to humans. The first level of development is defined by innate *instinctive responses*. The second level is characterised by the emergence of *conditioned reflexes* (or classical Pavlovian conditioning). The third level is defined by the emergence of *rational or intelligent behaviour* (as exemplified by the apparent ability for "invention" and "tool-use" in apes). The fourth and final level of development is reached by the emergence of auxiliary cultural symbol systems that human beings develop to enhance and restructure their natural cognitive abilities.

expected in the actual evolutionary process. Real evolutionary history is messy. Different levels of cognition could have been adopted in overlapping and non-uniform ways by different lineages in nature. Furthermore, as higher levels are adopted during evolution, lower level cognitive capacities will still be active in higher organisms alongside these new capacities. Keeping all these caveats in mind, let us now look at the details of Dennett's Tower of Generate-and-Test.

On the bottom level are *Darwinian Creatures*. These creatures are capable of only a very limited range of responses to environmental challenges. Their responses are completely due to hard-wired and naturally selected behavioural control mechanisms. Once the creatures are created they are "field-tested" in their environments by natural selection, so that only the best designs reproduce successfully. Representational capacities are non-existent at this stage, and learning does not appear at all.

It might be said of these creatures, although only in a metaphorical sense, that they "generate and test" hypotheses about their behaviour by employing their hard-wired mechanisms. But it is actually natural selection that "generated" and "tested" the hypotheses over successive generations, keeping those that were successful in promoting the survival of organisms and rejecting those that were detrimental.

Eventually in Dennett's model the bottom level gives rise to creatures with a new design feature: *phenotypic plasticity*. These creatures' phenotype is not completely fixed at birth, but retains the capacity throughout the organism's lifetime to be adjusted by environmental contingencies. For the first time, then, creatures are endowed with learning mechanisms that enable them to try out different behavioural options in their environments. But even with this improvement, Dennett thinks, some of these creatures are no better off than the hard-wired ones, because they lack the ability to favour smart behaviours. Nothing within the organisms tells them which behaviours are smart and therefore worth pursuing over less smart ones.

Other organisms are more fortunate, and have wired-in reinforcers and punishers that favour the smart behaviours. These fortunate ones are much more flexible in their encounters with the environment: they interact with their environment by generating a variety of behaviours, which they try out, one by one, until they find a response that actually works. And they "know" that it works because they receive a positive or negative signal from the environment, which adjusts the probability of that response's being reproduced again on similar occasions in the future. Some of these creatures are unfortunate enough to have their reinforcers wired-up wrongly, so that they get a positive signal for responses that are actually detrimental to the organism. As a consequence such organisms quickly vanish from the environment. Only a lucky few are born with the reinforcers wired-up correctly, and this gives them a huge advantage over the bottom level.

As a consequence of this design improvement, a new kind of creature emerges on the scene: the *Skinnerian Creature*. But, Dennett reminds us, this new level relies heavily on creatures not making any fatal errors when engaged in their interaction with the environment. Such creatures randomly produce behaviour after behaviour, and hope that no single behaviour will bring them premature death. Their only hope is that an advantageous behaviour crops up in good time and that this behaviour will be retained through reinforcement for the future. Still, these are the first creatures that can learn from the environment in classical and instrumental ways. Some remarks on these types of learning are in order here.

The basic learning Skinnerian creatures manifest is typically underwritten by the formation of associative links between two or more units (mental ideas, neural activity, or observable behaviour). David Hume (1973), for instance, thought that all thinking is mediated by a set of innate associative principles that link two or more (mental) ideas together by their similarity, contiguity, or causal succession. However, classical behavioural scientists such as C.L. Hull (1943), B. F. Skinner (1938), and E.L. Thorndike (1911), have been reluctant to postulate internal representations in the case of animals. For them, an association is a (learned) connection between two or more units of neural activity within the nervous system. These connections between sets of neurons are formed either because two or more units have often been active at the same time, as in classical or Pavlovian conditioning (Pavlov, 1927), or because the connection between two or more units have been strengthened by reward or punishment, as in operant or instrumental conditioning. C.L. Hull (1943), for instance, believed that neural activity caused by the triggering of the conditioned stimulus (such as a tone) acquires the ability to activate the motor muscles whose contraction leads to the observable response (salivating) in case the response is rewarded each time it follows the conditioned stimulus. The strength of the associative link depends on a set of features such as delay of reinforcement, how many times the stimuli have been paired together in conditioning trials, and the saliency of the respective stimuli.

Dennett is clearly thinking of this tradition in animal psychology when he introduces his Skinnerian Creatures. He writes,

> The cognitive revolution that emerged in the 1970s ousted behaviorism from its dominant position in psychology, and ever since there has been a tendency to underestimate the power of Skinnerian conditioning (or its variations) to shape the behavioural competence of organisms into highly adaptive and discerning structures. The flourishing work on neural networks and 'connectionism' in the 1990s, however, has demonstrated anew the often surprising virtuosity of simple networks that begin life more or less randomly wired and then have their connections adjusted by a simple sort of 'experience' – the history of reinforcement they encounter. (Dennett, 1996, p. 85)

Clearly, then, the second level of the Tower does not provide sufficient evidence for rich internal representational systems that guide the animals in their decisions to act. While it is true that these animals are creatures that learn from environmental contingencies, their learning is nevertheless nothing more than the formation of conditional reflexes. Given the right stimulus configuration, they produce the appropriate response, a response that was previously reinforced through reward in connection with this kind of stimulus.

As with Darwinian creatures, one can ascribe to Skinnerian creatures the ability to generate and test hypotheses about their behaviour. For the learning mechanisms endow the organism with the ability to generate hypotheses about its behaviour in a random fashion, and they also allow the organism to test its hypotheses in terms of reward and punishment. The organism's abilities, however, are highly limited. For the original hypotheses are only randomly generated, and, moreover, they are selected by an innate faculty of reward and punishment. This notwithstanding, however, learning mechanisms give the Skinnerian creatures a huge advantage over the Darwinian ones in terms of behavioural flexibility.

As Dennett observes, Skinnerian-like conditioning procedures are good as long as the creatures are not killed off by one of their early errors. A much more efficient solution would involve a kind of internal pre-selection among all the possible alternative responses, so that the erroneous responses can be selected out even before the creature actually performs them in the real external environment.

An ability for pre-selection implies the existence of an internal model or representation of the world in which pre-selection can occur. The internal model or representation of the world – "an inner something-or-other" or an "inner environment" as Dennett calls it – functions like a filter that favours only those surrogate actions that are more likely than not to be the actions that the real environment would also favour. Whether this representation of the world is achieved within the animal by innate specification or acquisition through learning, the information must be coded in such a format that it can allow for the pre-selective effects mentioned above to occur.

Quoting Karl Popper, Dennett describes the new design feature as permitting "our hypotheses to die in our stead", and he calls creatures exhibiting this design feature *Popperian Creatures*. Popperian creatures do not have to act randomly. They can evaluate the consequences of moves prior to their being acted out as a result of their inner model of the world. It is this third level of the Tower of

Generate-and-Test that finally gives rise to the first stage of representational capacities in animals.

Dennett discusses the example of "latent learning" in rats to clarify what he has in mind. E.C. Tolman (1932; 1948) has shown that rats left to explore an unknown maze in which no food or other rewards are presented are still able to (latently) learn their way around the maze. For if on a subsequent test trial a reward is introduced into the maze, rats that had been previously exposed to the maze perform consistently better in finding the reward than rats which see the maze for the first time. It seems, then, that pre-exposure to the maze enables the rats to build a *cognitive map* of their surroundings even though they are not specifically rewarded for doing so. In as far as they construct models of their environments (=cognitive maps) rats belong to the category of Popperian creatures. Their models allow the rats to pre-select profitable actions (= routes in a maze) even before they actually act them out in their environment. Dennett contrasts this particular ability to construct *cognitive maps* with the "simpler" Skinnerian-like conditioning procedures such as classical and operant conditioning. Whereas the Skinnerian-like procedures do not require representational abilities, the Popperian-like procedures do.

From the discussion of cognitive maps above, it is evident that Popperian creatures employ sophisticated mechanisms to generate and test hypotheses about their behaviour. First, the hypotheses are not generated in a random fashion, they are guided instead by representations of the environment. And, second, the testing of hypotheses becomes possible even prior to their enactment in the environment. And, last but not least, Popperian creatures are the first creatures that demonstrate conclusive evidence of representational capacities underlying behaviour.

On the fourth and upper most level of the Tower are *Gregorian Creatures*. Those are creatures that have inner environments like Popperian creatures. But their inner environments are capable of being "informed by the *designed* portions of the outer environment". Those designed elements include tools, institutions, and symbolic codes; this marks the point of origin of culturally sophisticated organisms. Dennett explains this ability with findings of the psychologist Richard Gregory.

Like Gregory (1981), Dennett observes that, for instance, a pair of scissors is not only a result of intelligent design, but also an endower of intelligence. It is a kind of potential intelligence. If you give someone a pair of scissors, you will thereby enhance the potential to arrive more efficiently at the correct solution to a given problem. Artefacts, and other designed tools, thus have two aspects. Not only does their recognition and production require intelligence, but they are also able to confer intelligence on those who are using them. At this final level of representational capacities, therefore, the mind is enriched by culture and traditions. Human beings can profit from a rich system of cultural representations, and, most prominent among them, words of their languages. Words are the ideal mind-tools that can spread easily in communication from individual to individual and that can bring with them knowledge and solutions to problems that have been acquired through hard work over long periods of time. By being immersed within its own culture, the individual participates in the amassed wisdom of its cultural heritage.

Gregorian creatures are thus the creatures with the most powerful mechanisms to generate and test hypotheses about their behaviour. Not only can they generate and test hypotheses dependent on their own limited experience, but they can also generate and test hypotheses dependent on a principally unlimited experience of other organisms present and past. Moreover, the Gregorian creatures' inner environment is not only open to the representation of natural objects, but also to the representation of artificial or cultural objects with all their implications of potential intelligence. All these features taken together endow Gregorian creatures with an unparalleled cognitive sophistication that allows them to adapt their behaviour flexibly to previously unseen degrees and levels.

1.2 Problems for the Tower of Generate-and-Test

Dennett presents his Tower of Generate-and-Test as a plausible hypothesis about the evolutionary stages of cognitive capacities. But the consistency of his model is threatened by a major confusion within his larger theory of mental representation and its relation to the Tower. The confusion is most virulent in Dennett's attempt to draw an exact line that separates *unthinking* from *thinking* creatures. The source of the confusion has its root in a deeper ambiguity about the nature of mental representation. In the following I shall discuss Dennett's broader theory of mental representation in so far as it is relevant for my argument.

For Dennett, it is only with the appearance of mental representations that thinking creatures appear on the scene. Dennett (1996) believes that mental representations must satisfy two conditions:

(1) Mental representations are meta-representations and thus belong to (at least) second-order intentional systems.

(2) Mental representations are explicitly encoded within cognitive systems and are thus able to be manipulated by internal operations. They thus cannot be merely implicitly embodied within the design of the cognitive system.

It is the simultaneous requirement of these two conditions within the broader theory of mental representation and Dennett's use of representations in the Tower model that ultimately generates the confusion present in his analysis. But before engaging in the analysis of the confusion, I must discuss the two conditions in more detail.

(1) The first requirement of mental representations depends on Dennett's (1987) wider views on our techniques and strategies for understanding the world, and the role, which mental representations play in such strategies. According to Dennett, we use different

"stances" to explain and predict the behaviour of objects we encounter in our world.

The first stance we use is the physical stance. According to this, we explain and predict what happens around us by using our knowledge of physical laws. We explain, for instance, why a picture fell to the floor and broke by invoking our knowledge of gravity and the physical properties of glass. But most physical systems are extremely complex, and thus stretch our knowledge of the physical world too far: using the physical stance is therefore too impracticable. When that happens, we change in the first instance to the design stance.

The design stance allows us to treat objects as designed artefacts and thus to explain and predict their behaviour from the assumption that these objects are well designed to perform a specific function. So we can, for example, handle an alarm clock by pushing some buttons because we know that the clock is well designed to perform the function of waking us up at the specified time. But we do not have to know all of the physical details of how the alarm clock realises its function. As long as we know how it functions and what several of its parts do, we can sidestep all the physical details. But in some complex cases, again, even the design stance proves impracticable because the design of some objects stretches our knowledge too far. We can then use what Dennett calls the intentional stance, and it is with the intentional stance that mental representations come into play.

The intentional stance involves treating objects as rational agents. In an attempt to explain and predict their behaviour, we attribute beliefs, knowledge, desires, and goals to them. Once we assume that such objects are rational, we can then understand their behaviour as that which they ought to perform all things considered. A great benefit of the intentional stance is that we can apply it even if we do not know much about the internal organisation of the object to which we apply the stance. That is we do not have to know much about its internal design or its physical set-up. As long as we can assume that the object is rational, we can go on and explain and predict its behaviour. Anything that is explainable or predictable by the intentional stance is thereby an intentional system that has beliefs, knowledge, desires and goals about its world:

> any object [... whose behaviour is well predicted by this strategy is in the fullest sense of the word a believer. (Dennett, 1987, p. 15)

Intentional systems are objects that have representational states such as beliefs, knowledge, desires, and goals – as long as the strategy of attributing such states to them proves successful in explaining and predicting their behaviour. But this alone is not sufficient to properly *think* in the sense that Dennett envisions. Dennett (1996, p. 120) maintains that something like "the step from first-order to second-order" intentional systems is needed to create properly thinking organisms. A first-order intentional system, in this context, is a system that possesses representational states such as *beliefs* and *desires* about things, but does *not* possess representations of those representational states themselves: it is not capable of forming *meta-representations*. Dennett claims that unless a creature is capable of entertaining meta-representations it cannot properly entertain thoughts either. He has good reasons for supporting this claim.

Dennett's definition of an intentional system is always *relative to* someone using the intentional stance successfully to explain and predict the behaviour of the intentional system. The beliefs, desires, and goals of the intentional system towards which the stance is taken must therefore be conceived of as dependent on the role of an interpreter applying the intentional stance. The interpreter herself, moreover, needs to form representations about the beliefs, desires, and goals of the target system in order to apply the intentional stance towards it: she will thus have to form meta-representations about representational states of the object. So unless the interpreter is herself a second-order intentional system, she will not be capable of using the intentional stance towards others. As we have seen, Dennett believes that only with the acquirement of this stage will an organism manifest proper thinking. Mental states are therefore bound up with second-order intentional systems via their role in the intentional stance.

(2) The second requirement on mental representations is not that straightforward. Even though higher-order intentionality seems to be necessary for *thinking*, Dennett believes that higher-order intentionality is by no means sufficient for thinking to occur. There is an additional condition for thinking that must also be satisfied – mental representations must be *explicitly* encoded within the cognitive system.

To clarify his point that mere higher-order intentionality is not sufficient for the appeal to mental representation, Dennett discusses the work of the ethologist Carolyn Ristau (1991). Ristau's work on the "distraction display" of low-nesting birds is, Dennett claims, one of "the best studied examples" of apparent higher-order intentionality among animals. Predators, e.g., foxes, approaching the nests of low-nesting birds activate deceptive strategies in these birds such as the "broken-wing display". In these circumstances, the bird quietly moves away from the nest filled with eggs or nestlings and begins ostentatiously to feign that it has a broken wing: fluttering its wings, collapsing on the floor, and calling out in distress. This stage-play has the function of attracting the attention of the predator and of moving him away from the nest.

As Ristau (1991) found, individual birds manifest quite sophisticated control of their distraction display. For instance, they are able to sort approaching animals into different categories. Depending on the category detected, they modify their displays appropriately. So, for instance, if a cow approaches the nest, the bird displays only wing flapping to advertise its position and does not move away from its nest in order that the cow will not step on the nest. Furthermore, the birds seem to monitor the direction of predators' gaze and other behavioural parameters to modify their broken-wing displays in response to predators' attentional states. If they see that the predator loses attention, they modify their display and perform more vigorously to catch the predator's attention once again and to lead him away from the nest.

According to Dennett, however, this apparent higher-order intentionality still falls short of thinking. Dennett maintains that the rationale behind this behaviour is easily understood by putting it into an *imaginary* soliloquy of the following sort: "I could distract him if he thought that I was an easy meal. If I just give him enough behavioural evidence to manipulate his beliefs, I could motivate him to chase after me, etc." But this sort of soliloquy is just *imaginary*. The bird does not really go through these considerations. The soliloquy is what Dennett calls a *free-floating rationale* of the bird's behaviour. The meaning of a free-floating rationale is not exactly clear, but there are particular properties that Dennett thinks are necessary to the free-floating rationale.

In particular, Dennett takes a free-floating rationale to be one that is characterised by the following formulations that refer to the same thing:

> not a tactic the [organism] has figured out for itself, or been capable of reflecting on. (Dennett, 1996, p. 123)

> an elaborate rationale with "little or no evidence that the animals actually represent these rationales to themselves in any fashion." (Dennett, 1996, p. 124)

> honored in the designs of their nervous systems, but we are looking for something more; we are looking for rationales that are *represented* in those nervous systems. (Dennett, 1996, p. 131)

These passages make abundantly clear that what is missing in a *free-floating rationale* is some "explicit" representation that would enable the individual to access and manipulate this rationale in further reasoning processes. It is only on the basis of an explicit representation of something that was previously only implicitly or tacitly embodied within the system that an individual is empowered to understand and exploit such knowledge in his reasoning. As Dennett (1996, p. 133) notes,

> To anchor a free-floating rationale to an agent in the strong way, so that it is *the agent's own* reason, the agent must 'make' something. A representation of the reason must be composed, designed, edited, revised, manipulated, endorsed.

Such explicit representations give an individual greater flexibility, because now the organism can combine or integrate this representation with other representations stored in her working memory or other stores within her head. Thus, all things considered, apart from having the necessary feature of higher-order intentionality, thinking creatures also need explicit or symbolic representations.

But as soon as one tries to project those considerations about *mental representations* back to the model of the *Tower of Generate-and-Test*, one begins to sense a major confusion, which threatens the consistency of the model.

As we have seen, Dennett puts much emphasis on the need for "explicit representations" to anchor *free-floating rationales*. This requirement seems to point to the third level of the Tower of Generate-and-Test: the Popperian Creatures. These are the first creatures to demonstrate genuine evidence of representational capacities. Such creatures use their representational capacities to build representational models of the world that act as a filter through which organisms pre-select behaviour that is likely to further their ends. It would indeed be difficult to imagine how such pre-selection could work if the creatures could not represent the behaviour and its consequences. Such model building is the crucial design improvement that distinguishes Popperian creatures from Skinnerian creatures with their simple trial-and-error learning. Thus, one would expect that *thinking creatures* start with Popperian creatures at the third level of the Tower of Generate-and-Test.

Surprisingly, this is not what Dennett maintains. Dennett writes instead, that,

> The sort of higher-order sensitivity required to meet the needs of a piping plover or a hare or a gazelle can probably be provided by networks designed almost entirely by Darwinian mechanisms, abetted here and there by Skinnerian mechanisms. ABC learning [Dennett's term for associative learning], then, could probably suffice to produce such a sensitivity [...] these animals are Popperian creatures – creatures who can be guided by past experience to reject tempting but untested candidates for action – but still not explicit thinkers. (Dennett, 1996, p. 130-131)

> The free-floating rationales that explain the rudimentary higher-order intentionality of birds and hares – and even chimpanzees – are honored in the designs of their nervous systems, but we are looking for something more, we are looking for rationales that are *represented* in those nervous systems.[46] Although ABC learning can yield remarkably subtle and powerful discriminatory competences, [...] these competences tend to be anchored in the specific tissues that are modified by training. (Dennett, 1996, p. 131-132)

Because of Dennett's identification of "explicit representation" with thinking, and Dennett's refusal in the cited passages to regard Popperian creatures as thinking creatures, this level of the Tower of Generate-and-Test threatens to become inconsistent. As the above passages suggest, Dennett now wants to interpret Popperian creatures in terms of procedures of ABC learning.[47] ABC learning is Dennett's catch phrase for Associationism, Behaviourism, and Connectionism: all forms of learning that forge associative links between e.g. neural activity in neural networks.

When Dennett discussed the Popperian creatures earlier in his book, however, he mentioned Tolman's work on *cognitive maps* to distinguish explicitly Popperian creatures from simpler Skinnerian creatures and their ABC learning. Dennett writes,

> The behaviorists tried valiantly to accommodate latent learning into their ABC models. One of their most telling stopgaps was to postulate a 'curiosity

46 Note also that Dennett is prepared in this passage to explicitly grant the applicability of higher-order intentionality to birds and hares. This suggestion would posit that it is possible, at least in principle, to account for meta-representational abilities in terms of simple ABC learning procedures. But Dennett doesn't give any arguments that would warrant this conclusion.

47 Note that the same reasoning would also threaten to undermine Dennett's own example of *cognitive maps* in rats when discussing his Popperian creatures. Cognitive maps, and other spatial abilities, seem to employ a highly integratable representational system. A defining feature of cognitive maps is that they can enable the animals to plot out novel routes through their environments. These routes are novel in the sense that they have not been learned by any time-consuming trial-and-error learning. At least in these cases it is very difficult to see how such systems could perform without a minimal ability to cognitively integrate the spatial representations to find new routes.

> drive', which was satisfied by exploration. [...] As an attempt to save orthodox behaviorism, this move was manifestly vacuous. (Dennett, 1996, p. 92)

And shortly after this passage, Dennett again contrasts ABC learning with the "higher abilities" of Popperian creatures,

> We human beings are conditionable by ABC training, so we are Skinnerian creatures, but we are not *just* Skinnerian creatures. [...] We are Popperian creatures. (Dennett, 1996, p. 92)

And, in a final, passage, Dennett goes even so far as to level a classical objection to the behaviourist's methodology of ABC learning,

> Pigeons were Skinner's favorite experimental animals, and he and his followers developed the technology of operant conditioning to a very sophisticated level [...] Notoriously, the Skinnerians never succeeded in proving that pigeons were *not* Popperian creatures; and research on the host of different species, from octopuses to fish to mammals, strongly suggests that if there are any purely Skinnerian creatures, capable only of blind trial-and-error learning, they are to be found among the simple invertebrates. The huge sea slug [...] has more or less replaced the pigeon as the focus of attention among those who study the mechanisms of simple conditioning. (Dennett, 1996, p. 92)

So what is going on in these conflicting passages? There are two interpretations, one uncharitable, the other more charitable that can explain this conflict.

a) An uncharitable interpretation of Dennett's problem

The uncharitable interpretation assumes that Dennett is deeply confused about his two projects: the project of the Tower of Generate-and-Test and the project of a definition of thinking. The two projects seem to pull Dennett's intuitions in two completely different directions. Thinking about the evolution of representational capacities leads Dennett to focus precisely on the differences between ABC learning as understood in classical or operant conditioning terms or connectionist networks (=Skinnerian creatures) and the representational capacities of more sophisticated creatures with inner models of the world (=Popperian creatures). But looking at his favourite theory of *thinking* and its connection with *anchored rationales, meta-representation,* and *language* leads Dennett to suggest that the differences between Skinnerian and Popperian creatures are not differences in kind after all, but only differences of degree. This would explain the contradicting claims that Dennett makes about these two levels of the Tower of Generate-and-Test.

Dennett's confusion thus originates because Dennett tries to conjoin two incommensurable research interests. But the blurring of Skinnerian and Popperian creatures should be resisted when an evolutionary account of representational capacities is envisaged. As was clear in Dennett's own discussion of the Tower of Generate-and-Test, the Popperian creatures are distinguished from the Skinnerian creatures by their employment of representational abilities. It is their use of representational abilities that is crucial from an evolutionary perspective.

b) A charitable interpretation of Dennett's problem

The more charitable interpretation grants Dennett a reason for downplaying the differences between Skinnerian and Popperian creatures. It argues that the crucial distinction between thinking and unthinking creatures should not be between behaviouristic learning mechanisms and representational abilities generally, but more importantly between implicitly used and explicitly used representations. This interpretation of Dennett's problem will thus have to include a detailed analysis of what Dennett's understanding of the term "explicit representation" or "symbol" amounts to. A highly relevant passage for this purpose is the following:

> What would have to be added to the cuckoo's computational architecture for it to be able to appreciate, understand, and exploit the wisdom interwoven in its neural nets? A popular answer to this question, in its many guises, is 'symbols!' [...] How could it not be the case that implicit or tacit knowledge becomes explicit by being expressed or rendered in some medium of 'explicit' representation? Symbols, unlike the nodes woven into connectionist networks, are movable; they can be manipulated; they can be composed into

> larger structures, in which their contribution to the meaning of the whole can be a definite and generatable function of the structure – the syntactic structure – of the parts. (Dennett, 1996, p. 133)[48]

In this passage, Dennett identifies "explicit representations", the ones required to anchor the rationales of behaviour to the explicit control of individual organisms, with "symbols" that have a formal structure to which the operations of the system are sensitive.

By suggesting that animals only have access to "implicit" or "tacit" knowledge about the rationale of their behaviour – by suggesting, that is, that the rationales are only honoured in the design of their nervous system – Dennett claims that animals lack any genuine (explicit) representational capacities that enable them to exert control over their own behaviour. Animals are *not* able to "move", "manipulate", "compose", "integrate", or "transport" their knowledge or representations across different types of problems faced in their environment. Whatever types of representations animals employ in their "reasoning", they do not admit of *cognitive manipulation*. Such representations simply do not possess the correct formats to be so manipulated. As a consequence, Dennett does not think that non-human animals can really and truly think.

Proper representational abilities, and thus thinking, probably only emerge with the fourth and top-most level of the Tower of Generate-and-Test: the Gregorian creatures. And thinking emerges only for those Gregorian creatures, so Dennett argues, that have "initially 'stupid' practices [e.g., talking to themselves] – the mere mouthing of labels, in circumstances appropriate and inappropriate – that could soon turn into the habit of representing one's own states and activities to oneself in a new way." (Dennett, 1996, p. 149) Once a child learns how to use labels and acquires the habits of attaching those labels to situations she has experienced, she has created a new class of objects that can themselves become objects of further processing by older cognitive equipment (such as pattern-recognition processes). This will give the maturing child improving

48 The same passage also occurs in Dennett (1994).

access to and control over her representations through her new ability to process these re-representations (words) rather than only the representations themselves. Whatever knowledge or "representation" creatures of other levels enjoy, it isn't the right kind for thinking! Such cognitive structures may be called "representations", but they are only "tacitly" or "implicitly" appreciated by the animals and are not able to be cognitively manipulated and controlled. An explicit appreciation of representations implies the abilities of meta-representation and communication.

All these considerations seemingly combine to strengthen Dennett's thesis that animals do not have control over their own behaviour, since they cannot construct explicit representations. If animals have representations at all, they are not states accessible to the animal, which it can use. Now Dennett's thesis about behaveioural control and its concomitant explicit representations is surely an empirical issue. In the end, it is only empirical data that can demonstrate whether animals use explicit representations or not. It is thus precisely at this juncture that recent developments in animal psychology become highly relevant for answering the questions Dennett raises. I think that this evidence shows conclusively that Dennett is wrong in his conclusions about the lack of representational control of behaviour in animals. The cognitive revolution in animal psychology has led to major re-definitions of the underlying mechanisms mediating conditioning in animals. It is to these issues that I want to draw the attention to next.

2. An improved evolutionary taxonomy

In section 1. I discussed Dennett's model, and I showed how it is infected by a confusion resulting from his two projects of defining representation. Dennett's insistence on explicit representations leads him to downplay the crucial evolutionary difference between the representational capacities of Skinnerian and Popperian creatures: the latter are now seen as differing from the former only in matters of degree. Whatever representations Popperian creatures use in their reasoning, they are according to Dennett most probably implicit representations that do not allow the creatures to control their own behaviour. Proper thinking, however, must be conducted in explicit representations that allow the creatures to control their behaviour by manipulating their own representational states. As I indicated, the question of implicit versus explicit representation with respect to non-human animals is an empirical issue, and it has yet to be seen whether Dennett is right in his conjectures.

There are two goals I shall accomplish in the current section. First I shall clarify that there is indeed a crucial difference between the representational systems of Skinnerian and Popperian creatures and what that difference is exactly. Secondly I shall adduce empirical data to show that Dennett is wrong in supposing that animals are not capable of explicitly encoding representations in their reasoning. I shall proceed in a similar fashion to Dennett by outlining a stage model of the different capacities of nonhuman animals that incorporates the new information.[49]

49 I will not discuss Dennett's Darwinian creatures in my proposal since these creatures are uncontroversial. The controversy starts with creatures that are capable of learning about their environment. Here the debates centre on the question of whether or not relatively basic general learning abilities such as trial-and-error or instrumental learning can be underwritten by genuine representational states. Some scientists believe that there is not enough evidence to demonstrate that such basic general learning abilities need representations, others think they some such basic learning abilities need to be underwritten by genuine representational states. My proposal therefore focuses on creatures that are already capable of learning about their environment. I do not want to imply therefore that Darwinian creatures are not an important evolutionary stage for the development of cognitive capacities. But, since they are uncontroversial, I will not discuss them here.

2.1 Basic Learners

On the first level of the evolutionary stage model of cognitive capacities are what I shall call *Basic Learners* who can adjust their responses to environmental contingencies in classical and instrumental ways. These learners closely correspond to Dennett's Skinnerian creatures. In the following I want to draw out differences among Basic Learners that are not appropriately highlighted in Dennett's account of Skinnerian creatures, but which prove interesting from an evolutionary perspective.

To better understand the scope of Basic Learner's general learning abilities we need first to be clear on how classical and instrumental learning work. Both help the organism to adjust to the local contingencies of its environment by exploiting predictive relationships that often occur in nature between events. A specific type of vibration, for instance, might predict the imminent arrival of an organism's preferred prey. It would obviously benefit the organism if it could learn about such contingencies in its environment. There are two kinds of basic general learning abilities that have been thoroughly studied in animal psychology: classical and instrumental learning (Dickinson, 1980; Gallistel 1990; Pearce, 1987). In the following I will present each of these kinds of learning in enough detail to draw out important consequences for the current taxonomy.

(A) Classical Conditioning

Classical or Pavlovian conditioning is said to occur whenever a previously neutral stimulus, such as a tone, that normally is of little interest to the animal becomes associated with a previously existing response of the animal to the extent that the neutral stimulus will now evoke the response by itself. Ivan Pavlov (1927) was the first to show how these learning mechanisms operated through laboratory experiments. Obviously, animals show capacities for classical learning in the wild too, but laboratory studies can investigate the details of the learning abilities under controlled observations. In one of Pavlov's most influential experiments he exposed dogs to sequences of a tone followed by the delivery of food into a bowl. The dogs' response to the biologically relevant event of food delivery is salivation. As Pavlov incidentally observed, after some trials the dogs began to salivate in the presence of the initially neutral stimulus of the tone alone, thereby showing signs of anticipating the delivery of food. Why did the dogs learn this? The dogs' acquisition of this response (salivation) to a new stimulus increases the efficiency of the digestive system, for the dogs will now be prepared to eat even before the food is actually delivered to them. This will give a dog an advantage over other dogs that did not learn about this relationship between events, since its digestive system is already fully up and running in anticipation of the food.

Usually, there are four components that one needs to distinguish in such simple conditioning situations as above:

(1) The *unconditioned stimulus* (US) elicits a predictable response from the animal's behavioural repertoire prior to any learning or conditioning. (Food in the above example)
(2) The *unconditioned response* (UR) is predictably elicited by an unconditioned stimulus prior to any learning or conditioning. (Salivation in the above example)
(3) The *conditioned stimulus* (CS) is a neutral stimulus which does not initially predictably cause the UR, but which is paired with the US during learning. (Tone in the above example)
(4) The *conditioned response* (CR) is a predictable learned response to a conditioned stimulus that was previously paired with an unconditioned stimulus. (Salivation elicited by the Tone)

The conditioned response, in other words, is the product of learning abilities that attach the organism's innate responses to newly occurring situations, which the animal was not equipped to react to prior to learning (such as the tone in the above example).[50] This analysis of learning obviously implies that prior to any learning there already exists associative links between the US and the UR that have been installed in the animal by natural selection due to their obvious survival value for the animal. These associative links are universal or species-specific behavioural traits the organisms need not learn about.

There are, however, two different theories about classical conditioning based on how one conceives of the associative links produced by classical learning abilities (Dickinson, 1980; Pearce, 1987). The CS can activate the UR in both of the following ways:

1. Associations form first with the US, which then in turn activates the UR indirectly.
   [Tone(CS) ---> Food in Bowl(US) ---> Salivation(UR)]

2. Associations form directly with the UR and thus activate the UR directly.
   [Tone(CS) ---> Salivation (UR)]

The first form is known as the S-S theory of conditioning, the latter form as the S-R theory of conditioning.[51]

Learning thus consists ultimately in the formation of new stimulus-response (S-R) connections. What animals learn when exposed to conditioning schedules – e.g. a tone (CS) followed by the presentation of food (US) – is nothing more than a *new way* of reacting to initially neutral stimuli (CS) by *old means* (the UR).[52]

50 In a typical conditioning situation the CR and the UR may come to differ from each other in some subtle ways, but one must keep in mind that the learned response is basically the UR simply adjusted to the new stimulus (Holland, 1992).

51 For a detailed discussion of these models, see Dickinson (1980), Holland (1992), and Pearce (1987).

52 Conditioned reflexes can become quite complex when one takes into account that the spreading of innate reflexes may travel along transitive paths. In

Learning by classical conditioning is at root a process that facilitates the spreading of innate reflexes to new stimuli or situations. But this learning is also inherently passive in that it must wait until the environment brings the relevant contingencies to the organism. If the animal is deprived of the relevant experiences of the contingencies, there will be no other way to learn about "new" ways of responding to situations.

Most importantly for my evolutionary stage model, learning by classical conditioning does not provide sufficient evidence of representational capacities. Learning by classical conditioning clearly involves the perceptual detection of various features of the environment, otherwise animals could not learn about the presence of new signals of relevant biological events. But must it thereby involve representations? To determine this, we need to have at least a preliminary grasp of what representations are. I will not go into the details about the precise nature of representations here, but I shall outline an initial understanding of representations that I can use for the purposes of this section.

A representation, according to my definition, is whatever plays the functional role of a state in a representational system. I base my definition of representations on an influential account of representational systems from Randy Gallistel (1990; 1992) and Robert Cummins (1975; 1983). Both Gallistel and Cummins analyse representational systems in terms of a formal mapping (isomorphism) between two systems: the represent*ing* system and the represent*ed* system. Gallistel (1990, p. 16) defines the formal mapping as a correspondence that holds,

higher-order conditioning schedules, for instance, experimenters have shown that once animals learn a first conditioned response, they can spread that response to new CS that are associated with the older CS as in the following depiction:
CS1--------->UR
CS2--------->CS1
CS2--------->UR

> when there is a procedure that maps entities, relations, and operations in the represented system into entities, relations, and operations in the representing system in such a way that two or more entities within the represented system are related in a given way *if and only if* there is a corresponding relation between their representatives in the representing system.

The proposal thus maintains that a representational system qua representing system satisfies the following conditions:

(1) There is a systematic mapping from external states of the environment to internal states of the representational system whose function is to *detect* changes in the environment.
(2) There is a formal mapping between the operations defined over the internal states of the representational system and operations defined over the external states of the environment.
(3) Both formal relations in (1) and (2) are such that the operations of the represent*ing* system generate *valid anticipations* about the external states of the represent*ed* system (the environment) prior to experience.

So a representational system must not only possess states that detect changes in the environment, but it must also possess internal operations defined over such states. The internal operations manipulate the states to derive valid anticipations or predictions about its represented environment prior to experience. The formal mapping of the states and operations holding between the two systems holds the two systems in sync. For unless the representing system were constrained by the formal mapping between its internal operations and the (causal) operations of the environment, it might produce anticipations that prove highly inadequate to its environment – e.g. it might derive false anticipations or predictions.[53] In a

53 Note that this explanation of representational systems acts as an elucidation of what Dennett thinks that Popperian creatures employ in their reasoning to pre-select their actions.

final move, representations are identified with states of such representational systems. To have representations, then, an organism must not only possess states that detect changes in its environment, but it must also be able to generate predictions about its environment even prior to experience on the basis of the internal operations of its representational system. If the states of the organism do not satisfy these requirements, they cannot count as proper representations.

In the case of Basic Learners, however, it is difficult to discern the existence of representations. True, they possess states that detect features of their environment. As we have seen, classical learning depends on the perceptual detection of new stimuli that drive learning of new contingencies. But the internal states of Basic Learners do not seem to be capable of generating valid predictions on the basis of internal resources alone, prior to actual experience. Basic Learners must always be exposed to the pairings of conditioned stimuli with unconditioned responses to learn about any contingency holding between the two. In the absence of internal operations generating predictions, the states of Basic Learners cannot be called representations. I shall call these states *detectors* to distinguish them from proper representations. So even though these organisms have the ability to augment their stock of *detectors* of new conditioned stimuli, the detectors do not constitute a genuine representational system.

(B) Instrumental Conditioning

Basic Learners, however, are not confined to learning by classical conditioning. Some of these creatures are able to actively change their environments via instrumental learning in ways suitable to them. Instrumental learning is said to occur whenever the following two conditions are satisfied:

First, a perceived situation elicits a set of behavioural responses from the organism. These responses are typically generated completely at random in a trial-and-error fashion in which all sorts of responses are tried out in the environment.[54]

Second, the organism is equipped with a feedback mechanism that tells it which responses to retain and which to dismiss. If the outcome of a response is rewarding to the organism, the connection between the situation and the response strengthens: the probability of its occurrence in the same sorts of situations increases. Similarly, if the outcome of a response is adverse to the organism, the connection between this response and the situation diminishes in strength: the response is less likely to re-emerge in the same kinds of situations in the future.

While classical conditioning can work solely with innately specified reflexive responses, the same is not true of instrumental learning. An instrumental learner is principally able to introduce completely new responses from whichever source and is therefore not tied to innately specified responses. This allows, on the contrary, for the modification of voluntary responses on the basis of the aforementioned feedback mechanism. Both kinds of learning help to establish new associative links between contextual features and particular responses of the organism (Thorndike, 1911). But instrumental learners possess the advantage of more sophisticated learning abilities because they can actively interact with their environment and observe the outcomes of their interactions, while conditional learners are confined to recording passively the contingencies they encounter.

Yet Basic Learners using instrumental learning still do not provide sufficient evidence for the postulation of representations. That is, nothing so far forces us to claim that they must *represent* the outcomes of their responses rather then just detect them somehow. One argument against this might assume that organisms perform their responses because they *want* the particular outcomes. If that were true, it would follow that organisms would have to cognitively

54 Importantly, the set of responses might also include behaviour that is not previously contained in the innately specified behavioural repertoire of the organism.

represent the outcomes since the outcomes will not be present at the time of the response's performance. However, there are obvious alternative accounts of what happens in instrumental learning that have to be ruled out first before one can even begin to develop an convincing argument for representations. It may be the case for example that organisms perform their responses simply because they have learned in the past in a trial-and-error fashion to associate a specific response to contextual features. The associative link will be automatically formed simply on purely neuro-mechanical grounds whenever the response is followed by a rewarding outcome.

Instrumental learning thus can be realised as a simple feedback process that strengthens a response in the context of a perceived situation whenever it results in a positive outcome, and weakens a response whenever it results in a negative outcome. But the positive or the negative outcome of the response need not be explicitly represented in the organism at all. The same kinds of reasoning about representations that applied to classical learning will apply equally to instrumental learning. While it is true that instrumental learning implies the perceptual detection of the situations the organism finds itself in, the organisms do not seem to possess representational systems that are capable of deriving valid anticipations prior to experience. It is more plausible to say that instrumental learners generate responses in a trial-and-error fashion and let the environment select which ones are retained and which ones are disregarded for future uses.

To summarise the discussion of Basic Learners: The evidence about Basic Learners, both classical and instrumental ones, does not provide a conclusive case for the postulation of representational systems. On a very abstract level, Basic Learners can be viewed as *reflexive machines* that adjust their motor responses to impinging stimuli from the environment. It is not the internal resources of the organisms that choose their responses. Rather it is the environment that selects the appropriate responses either by presenting the organism with the relevant contingencies or by influencing the probabilities of the responses through the positive or negative outcomes.

Furthermore, Basic Learners are organisms that closely correspond to Dennett's Skinnerian creatures organisms. All these creatures rely upon powerful general learning abilities to interact with their environments. But they do not yet show signs of possessing representational systems.

2.2 First Representers

Eventually, evolutionary development gave rise to a new breed of more sophisticated *First Representers.* These organisms are among the first to use representational systems in their reasoning. As we have seen, the postulation of representational systems requires, at a minimum, not only internal states detecting features of the environment, but also internal operations that can manipulate those states. Psychologists wanting to postulate such representational systems in animals need therefore to present convincing data that cannot be explained by the kinds of general learning abilities used by Basic Learners. Interestingly, psychologists claim to have found such data. There is now strong consensus among animal psychologists that some animals are able to use inner representations to mediate even some of the mechanisms of general learning abilities. There is now a recognisable interest in the mental representations underlying some behaviour:

> Most students of animal conditioning today agree that the more profitable subjects of inquiry are the mental events, structures, and processes that underlie conditioned behavior, and not the conditioned behavior itself. Unlike in the heavy-handed behaviorisms of the not-too distant past, the domain of most modern conditioning theories is not one of spit and twitches, but a conceptual one. (Holland, 1992, p. 106)

Surprisingly, however, this is not the impression one gets when reading philosophers who refer in their theorising to the animal conditioning literature. Proper conceptual thinking is still contrasted with processes of animal conditioning where such conditioning is seen as an instance of a brute and mechanical way of responding to the environment: basically spit and twitches. Among other things, the absence of a positive appraisal of animal conditioning may still be due to the lingering influence of the "heavy-handed behaviourism" mentioned in Holland's citation. In the following I shall present the crucial evidence that brought about the revaluation of animal cognition in animal psychology. Of course, I cannot here present all the evidence for this claim, so I will select the most influential experiments.[55]

How could we decide between theories that explain animal behaviour in "mechanical" ways and theories that explain animal behaviour by postulating representational systems? One way to decide between the competing theories is to find data that support the predictions of one theory, but reject the predictions of the other. So let us concentrate on a simple stimulus-response connection and let us draw out the predictions the two theories make about it. This stimulus-response connection is depicted as follows:

CS (tone)------------US (food)----------UR (salivation)

Classical conditioning theory claims that the stimulus (CS) gradually comes to control the response (UR) whenever the animal is exposed to the contingency and the response is beneficial to the animal. The forming of associative connections takes time and develops gradually, and can happen only if animals are repeatedly exposed to the connection. If the contingency should change as a consequence of the devaluation of the US, however, the response would lose its beneficial effects. In that case, the animal would have to learn about this change in its environment. But as all learning necessarily takes place over time, animals would stubbornly continue to display the UR in the presence of the CS at least for the first couple of trials immediately following the change.

55 For detailed reviews of the evidence see Dickinson (1980), Gallistel (1992), Pearce (1987; 1997), and Shettleworth (2001).

Cognitive theories predict quite a different outcome. If the animal really constructs representations of the CS and US during conditioning trials, it is the representation of US that directly controls the response of the animal and not the contextual features of the situation. Again, if the contingency should change as a consequence of the devaluation of US, what would happen to the response? Well a devaluation of the US would have an immediate effect on the representation of it: the desirability of the US would be updated by the representation. So once the representation is updated, and given the perception of the CS, the animal should abstain from manifesting the response from the first trial.

We thus have two differing predictions, and we need experiments to decide between the two. Holland and Straub (1979) carried out an influential experiment that does just that.

In a first phase of their experiment they exposed rats in a conditioning chamber to a noise (CS) that was followed by the delivery of food pellets (US). As is typical in such circumstances, after a number of training trials the rats start to approach (UR) the dispensing magazine in anticipation of the food. Rats are now motivated to approach the magazine site as soon as they perceive the noise.

In a second phase of their experiment, rats in the experimental group (EG) were allowed to access freely the food pellets for 5 minutes on each of the next 4 days in the absence of the noise. However, after consumption of the food pellets, rats were injected with a sickness inducing substance. The purpose of the injection was to condition aversion to the food pellets. Rats in the control group (CG) were subjected to a similar treatment, except that the injection and the free access to the food pellets were presented on separate days so that the consumption of the food pellets was unaffected by the induced nausea.

In a third phase of their experiment, the actual test, both groups of rats were returned into the conditioning chamber, where the noise was again presented, but this time without the presentation of food (in extinction). The all-important question Holland and Straub asked is whether there is a significant difference in the response rates of approaching the magazine during the tone (UR) between the two groups of rats. The results were conclusive. Rats of (EG) showed a significant reluctance to approach the magazine during the onset of the tone compared with rats of (CG).

Now if the rats had learned about the contingency by means of simple associative learning principles, the prediction would be that the EG rats should have approached the magazine with the same rate upon the onset of the tone. For the only way in which an already instantiated stimulus-response connection can be altered is by learning that the stimulus now leads to an unpleasant response. However, the rats were not exposed to the stimulus (the tone) during their free access to food in phase 2 of the experiment. Consequently, there was no possibility for the rats to modify the previously instantiated stimulus-response connection. That is also why, according to simple conditioning theory, the old response should occur as soon as the tone sets in again.

The EG's significant reluctance to approach the magazine during the onset of the tone in the test condition is therefore strong evidence against simple conditioning theory. The data show that rats have actually learnt something involving the representation of the food pellets in phase 2. Once it is assumed that rats construct representations of the food pellet during their learning trials in phase one, the outcome of the experiment is easily interpreted.

Most importantly for my purposes, Holland and Straub's (1979) experiment supply evidence for the postulation of a functioning representational system in rats. Not only do rats use internal states that represent features of their environment (a tone and a food type) and their responses, but they also use internal operations that manipulate the representations to derive valid anticipations about their environment prior to experience. Somehow the rats managed to combine or integrate their representations of the tone with their altered representations of the desirability of the food pellets to change their responses (magazine approach) without the benefit of further experience of the delivery of the food items in the test condition. They could not have done this if they were limited to the use of simple conditioning principles. We could summarise the rat's reasoning as follows:

1. Hearing the tone, the rat forms a representation of the tone.
2. Seeing the food, it forms a representation of the food.
3. Tasting the food, it forms a positive goal representation for the food.
4. Because of the set-up of the experiment, the rat learns about the contingency and therefore forms the representation that the tone (causally) accompanies the food.

As a consequence of these representations and their integration, as soon as the rat perceives the tone, the rat tokens the representation of food and ascribes a positive motivational quality to it. In other words, the rat expects that the food will be dispensed any moment.

5. The rat already possesses a representation that if valued food is present, then it should approach the site where the food is presented.

Putting the last representation together with the others, the rat will rush to the food site as soon as it hears the tone – even before the real food is dispensed because the whole process is now mediated by mental representations.

In phase 2 of the experiment, however, its goal state is devalued, and this actually blocks the integration of the goal representation with representation 5. – because now the goal representation of the food carries a negative quality. As a consequence, rats will not approach the magazine site on the test condition.

The cognitive powers of the representational system evidently depend on the specific ways in which the internal operations combine representations. One expects that the cognitive powers of a representational system will be directly proportional to the complexity of the internal operations that manipulate the representations. So how weak or strong are the internal operations of rats? Evidence suggests that they are not very strong, so whatever mechanisms realise the reasoning above will be susceptible to several shortcomings.

In a further experiment with rats, Rizley and Rescorla (1972) tested the nature of the rats' internal operations.

In phase 1 of the experiment, rats are exposed to a tone followed by a shock. As above, rats construct a connection between the representation of the tone and the representation of the shock in the training trials. The connection between these two representations leads to the acquisition of a conditioned fear reaction upon hearing the onset of the tone.

In phase 2 of the experiment, rats are exposed to a further contingency holding between a light and the previous tone. However, no subsequent shock is administered in this phase. Again, rats build a connection between the representation of the light and the tone on the basis of the training trials.

In phase 3 of the experiment, rats are exposed to the tone alone without the administration of the shock (in extinction). This training trial should have changed the previous connection holding between the representation of the tone and the shock, actually leading to the disappearance of the conditioned fear reaction when presented with the tone.

In the final test condition of the experiment, rats are again exposed to the onset of the light. The question is: How do rats respond to the light alone? Can their response tell us something about their internal operations?

To simplify the understanding of the experiment it might help to re-describe the experiment as a set of conditionals:

Phase 1:	If *tone*, then *shock*
Phase 2:	If *light*, then *tone*
Phase 3:	If *tone*, then *nothing*
Test:	If *light*, then what?

If we were allowed to assume that rats use rational capacities in anything like we do, we could easily predict their behaviour in the experiment above. Given the first two phases, a rational being would have combined the conditionals to derive the conditional that if the light goes on, then a shock will be administered – and we know that rats at least can combine such conditionals and would consequently show fear reactions as soon as the light is turned on. However, given phase 3 of the experiment, a rational being would have to reconsider its knowledge gained in phase 1 in the face of new evidence in phase 3 that contradicts phase 1. Once rational beings have updated their relevant memory stores, they would derive the conditional that if the light sets in, nothing will happen since the tone does not act as an indicator of shock any more. So once tested with light only, rats should not respond with any conditioned fear reactions to the light. But how did the rats actually respond? Contrary to the prediction above, rats responded with a very significant conditioned fear reaction to the onset of the light due to their previous learning of the tone-shock connection.[56] There are at least two possible ways to interpret this result.

56 The situation is actually more complicated than that. Rashotte (1977) performed a rather similar experiment with pigeons in which the respective conditionals took the following form:
If *white light*, then *food*
If *blue light*, then *white light*
If *white light*, then *nothing* (extinction trial)
If *blue light*, then?

In stark contrast to Rizley and Rescorla (1972), pigeons were seemingly able to integrate these conditionals. The *rational* point of view predicts that the pigeons should refrain from responding to the blue lights. This time the experimental outcome concurred with the rational prediction: the pigeons actually responded less frequently than the control group who was not exposed to the extinction trial. Now, this does not show that pigeons are either *true logicians* or better than rats – there is ample evidence in the animal learning literature suggesting that rats and pigeons generally employ the same learning mechanisms. Rather than this, the contradictory evidence suggests that there is something going wrong with the rats' mediating mechanisms in these experimental settings.

(1) One interpretation gives rats the benefit of rationality. The rats' observable behaviour is not a failure in rational reasoning at all, but does actually adhere to a rational principle of prudence. Under the circumstances of the experiment, a rat could have reasoned that since the tone is sometimes followed by the shock and sometimes not, it should be afraid whenever it hears the tone or whenever it anticipates the representation of the tone as it detects the light. Rats might reason that it is better to be safe than sorry in the long run. Their behaviour on the experiment thus does not conclusively count against rational reasoning.

(2) The other interpretation views the data as showing that rats are not able to reason rationally. This reading of the data, and one that comes naturally to animal psychologists, is that rats use an altogether different reasoning capacity supported by associative principles. Note, that contrary to classic behaviourism, however, the associative principles are now directly applied to the representations of organisms. The particular failures that animals show when tested on complex learning tasks might therefore be due to the nature of the associative principles animals use in their reasoning. It may be the case, for instance, that their reasoning is limited by performance limitations: the experiment may overtax the rats reasoning resources, or the experiment may blow the biologically relevant constraints on the rats' memory system, etc. In any case, such limitations would show up in the rats' failure to update information that would help the animal to find the correct solution to a problem.

Current evidence in animal psychology makes the second interpretation more credible. Holland (1979), for instance, showed in an influential experiment how precarious rats' reasoning is:

In phase 1 of the experiment rats are exposed to a contingency of a tone followed by the delivery of food pellets at a magazine site. As is typical for such experiments, rats quickly learn to approach the magazine site upon hearing the tone.

In phase 2 of the experiment, Holland changed the conditioning schedule into an omission schedule: now the delivery of food pellets is actually omitted whenever rats approach the magazine site during the onset of the tone. How do rats cope with the situation?

Any minimally rational being should refrain from approaching the magazine site as soon as it learned that approaching the magazine site during the onset of the tone would not lead to the satisfaction of its cravings for pellets. Surprisingly, though, rats persist in their magazine approach even under the omission schedule and thus insensibly forfeit their chances of easily satisfying their cravings. It is not completely clear why rats cannot change their behaviour to adjust to the omission schedule. One possible analysis, and also one that will throw more light on the issue of animal reasoning, would hold that rats cannot inhibit their magazine approach once they have acquired it.

In accordance with the above, let us assume that the rats form a connection between the representation of the tone and the food. Once a rat perceives the tone, its representation is activated, and as a consequence of the established connection, this activation spreads and in turn activates the representation of food. This will give rise to the rat's expectation that food is imminent. Moreover, let us assume that the response of approaching a site where food is expected is a hard-wired response that is automatically triggered whenever a representation of food is suitably activated. Now, as long as the magazine approach leads to food intake everything is fine. In the omission schedule, however, rats never acquire food when they approach the magazine during the sounding of the tone. But the connection between the representation of the tone and the food is still there as a remnant of phase 1 of the experiment. So every time a rat perceives the tone, its representation of the food, and therefore its hard-wired approach response, will still be activated. Unfortu-nately for the rat, it cannot modify this hard-wired connection: in the instances in which the rat does not approach the magazine, it will still perceive the contingency between tone and food. Because of these circumstances, the rat cannot inhibit its approaching the magazine. I do not know if my analysis of the rat's observed behaviour is actually correct. But whatever the reasons that underlie the rat's failure, such failure shows that rats cannot freely use rational capacities to solve the problems encountered in the experiment. Their reasoning system seems to be beset by hard-wired connections that limit its flexibility. Whatever the rat's reasoning capacities turn out to be, and there is as yet no clear understanding of them, they will contrast sharply with rational procedures of reasoning.

To summarise: First Representers are the first organisms to use representational systems in their reasoning. However, their reasoning systems are incapable of using internal operations that obey anything approximating rational principles. Evidence from animal psychology suggests that First Representers may instead be using internal operations that obey some kinds of associative principles that have to work under strict resource limitations.

The discussion of First Representers has two important corollaries for the taxonomic project: it shows that (1) Dennett is wrong in thinking that animals are not capable of generating explicit representations which they can use to control their behaviour, and it also shows that (2) Dennett has, in his model of Generate and Test, overlooked organisms which use associative principles but can nevertheless use representational abilities in their cognition. Let me spell out each corollary in more detail.

(1) Recall that according to Dennett animals are not capable of using *explicit representations* that allow them to manipulate internal states and to control their behaviour on the basis of them. Whatever "representations" they may use, Dennett thinks that these can only be implicitly embodied in the design of organisms. The discussion of First Representers, however, has demonstrated that animals in fact can use explicit representations in their cognition. Such animals possess proper representational systems that can manipulate internal states by internal operations to derive valid anticipations prior to experience. I have adduced some experiments that demonstrate this thesis. True, those representational systems may not be particularly

sophisticated ones: they are susceptible to frequent breakdown due to the presence of hard-wired connections, and to several other biological limitations. But they still qualify as employing explicit representations that can be moved, manipulated, and composed into larger units that help the organism to derive predictions about its environment. The conclusion to draw from this discussion, then, is that Dennett's conjecture about explicit representations in animals is false. An improved evolutionary taxonomy of cognition, pace Dennett, needs to make room for the recognition of such types of organisms.

(2) One feature of Dennett's model is that it places all organisms that are capable of using only associative principles into the category of Skinnerian creatures. Moreover, Dennett sharply contrasts Skinnerian creatures with Popperian creatures on account of his belief that only the latter are able to use representational capacities in their reasoning. Now the discussion of First Representers has shown that the use of associative principles cannot be the right dimension to distinguish organisms that use representational capacities from those organisms that do not. Drawing on recent findings of animal psychology, I have shown that there are organisms that possess representational systems, but which can also use only associative principles to manipulate internal states. Dennett consequently wrongly categorises organisms as a result of his incorrect understanding of associative principles. For, as it turns out, there exist organisms with representational capacities that involve associative principles. This might be a small problem that Dennett might try to account for in his model, but it is a big step for the evolutionary development of cognitive capacities.

Because of all these complications, it is not easy to locate precisely First Representers in the framework provided by Dennett's Tower model. They are organisms located somewhere between his Skinnerian and Popperian creatures, depending on how one evaluates the cognitive processes of the organisms. The best way to deal with this problem is to introduce a separate category (from the

Skinnerian creatures) that recognises the different and special status of First Representers.

Once First Representers are recognised as a separate category, their use of representational systems permits us to draw further interesting implications about the structure of these representational systems. Spelling out these implications allows us to connect issues of this chapter with issues about cognitive evolution elaborated in earlier chapters. There we found that evolution manifests a tendency for specialisation in the physical and the cognitive realm. As we said in chapter I, we should expect, as a rule, that when two adaptive problems have solutions that are different, a single solution to them will be inferior to two specialised solutions (Cosmides and Tooby, 1994). Evolutionary psychology already familiarised us with the thesis that many species face complex environments with different adaptive problems, and we should therefore envisage the possibility that evolution fashioned organisms with compartmentalized minds. Once simple First Representers had ventured into more and more complex lifestyles, strong selection pressures on the basis of the trend to specialisation would have led to the formation of multiple cognitive modules that supported their new lifestyles. We can imagine how evolution would have continued this trend and how it would have consequently led species along a path of more and more diversified cognitive architectures. Moreover, we can further speculate that at a particular time in this progression, the degree of cognitive specialisation became itself a kind of a problem for the survival of the species. The complex cognitive architectures of some species would have created problems of modular interaction of the kind analysed in chapter III.

With this final state we see how one of the sufficient conditions for the development of general-purpose cognitive mechanisms outlined in chapter III is satisfied. But First Representers are not the most sophisticated organisms that evolution had fashioned. It is to these more cognitively sophisticated organisms that I next turn.

2.3 *Rational Reasoners*

Recent exciting evidence in animal psychology suggests, as I have said above, that evolution might have given rise to creatures with even more powerful representational capacities than the ones used by First Representers. Evolution clearly did this in the case of humans, but I am claiming also that evolution generated these powerful representational capacities in the case of non-human animals. Explaining what I take to be the nature of these new organisms, *Rational Reasoners*, is the topic of this section.

From our earlier discussion, we already know the kinds of demands the postulation of representational systems places upon the organism: First, organisms have to have representational states that are about features of their environments. Second, organisms have to employ internal operations that manipulate those representations to derive valid anticipations prior to experience.

The major design improvement of Rational Reasoners consists in the types of internal operations and representations they use. Some animals, that is, are seemingly capable of using rational principles in their reasoning about their environment. Now rational principles strongly contrast with associative principles on several counts. The most important one consists in the sensitivity of rational principles to the semantic content of the states they manipulate. An explanation of the last claim is in order here.

Reasoning by associative principles depends on the causal pairing of representations of the stimuli. Such causal pairing will be successful only if the paired stimuli satisfy features such as strength, timing, and frequency of pairings between stimuli: the unconditioned stimulus (US) needs to be strong or salient enough for the conditioning of the neutral stimulus (CS) to occur, the pairing of the US and the CS must be close enough temporally for the two to be associated together, and typically the US and the CS need to be paired frequently for the association to result in learning. If these features are satisfied, and so long as they are, learning will generally occur whatever the informational content of the representations associated during such learning. Put differently, it does not matter what the representations are representations of. Associative principles are predominantly sensitive to the statistical regularities of the representations. But as we saw in the previous section, First Representers also appear to be sensitive to structural features of the representations when they integrate representations to derive anticipations prior to experience. Now these derivations of valid anticipations cannot be explained completely on the basis of statistical information, since the animals are sometimes capable of integrating their representations to derive anticipations without the benefit of prior experience, and hence also statistical information. But as we have seen, the nature of these structural operations is not very well understood. Sometimes, these operations break down in the most unpredictable ways because of biologically relevant limitations. However, the states of First Representers do satisfy the definition of representational systems.

Rational principles, on the other hand, are highly sensitive to the contents of the representations even to the extent of discarding statistical regularities between them. The paradigm of rational reasoning can be illustrated through a typical example of the practical syllogism:

(1) The agent desires to drink some tea.
(2) The agent believes that by gripping the cup in front of it, it can satisfy its desire for tea.
(3) Therefore, the agent grips the cup in front of it.

Obviously, things might go wrong and consequently the agent will not be able to satisfy its desire for tea. It might, for instance, be that stronger desires overrule the agent's current desire or that the world does not play along and the cup is empty or filled with an undesirable liquid, etc.

A slightly more formal illustration of the above pattern of reasoning clarifies the all-important relation between the contents of the various mental states involved and the behaviour caused by them:

1. X desires B.
2. X believes that A leads to B.
3. Therefore, X performs A.

What these considerations suggest is that for behaviour to count as rational or goal-directed, it must be caused by cognitive states the contents of which relate appropriately, i.e. as shown in the practical syllogism, to the specification of the behaviour. The required causation is neatly captured by two constraints pertaining to the contents of the agent's desires and beliefs.

(1) The Instrumental Belief Constraint (IBC):
The agent would not have performed behaviour A, if it had not believed that A leads to the goal B.

(2) The Goal Constraint (GC):
The agent would not have performed the behaviour A, if it had not desired the goal B.

To count as a rational or properly goal-directed behaviour, behaviour A must satisfy both IBC and GC simultaneously. If either IBC or GC fails, but behaviour A is nevertheless performed, the behaviour cannot be interpreted as properly goal-directed.

Anthony Dickinson and his collaborators defend the view that animals are capable of performing rational or goal-directed actions (Dickinson, 2001; Dickinson and Balleine, 1993, 1994, 1999; Heyes and Dickinson, 1990). Dickinson et al. argue that rats' actions are guided by instrumental beliefs about causal contingencies and by goals that interact with beliefs in the ways outlined in the practical syllogism. Heyes and Dickinson (1990), for instance, base their arguments on the two constraints above by making them amenable to experimental manipulation.

IBC requires that an organism's behaviour be sensitive to changes in the instrumental beliefs and the environmental contingencies supporting them. We already know that rats and other animals are capable of learning about environmental contingencies and of encoding their learning in instrumental beliefs. Rats, for instance, are easily trained to lever press for the delivery of food pellets or sucrose solutions. Pigeons also are easily taught to peck at colour keys to access grains, etc. Moreover, it is well known, for instance, that once rats establish an action such as lever pressing for food, they are highly sensitive to changes in the causal contingencies underlying their instrumental beliefs (Dickinson, 1980; Pearce, 1987, 1997). Rats will thus stop, under the right circumstances, pressing the lever if food is delivered independently of the lever pressing or if lever pressing actually causes the omission of food (Davies and Bitterman, 1971).

But even though the satisfaction of IBC is necessary for behaviour to count as rational or goal-directed, it is not sufficient on its own to mandate such ascription. A rational behaviour must also satisfy GC. GC establishes that the behaviour must also be guided by a goal. We should always be wary about counting behaviour as goal-directed that occurs independently of any plausible goals the animal might have.

A good way to test the goal dependency of behaviour, Dickinson and Balleine (1993, 1994, 1999) propose, is to conduct a goal-revaluation test. This test can be illustrated as follows. Once animals are trained, for instance, to press a lever for valued food they also form the goal of accessing the valued food. To investigate whether the goal is an effective cause of the behaviour, psychologists attempt to devalue the goal by implementing a food aversion procedure. In such aversion procedures, rats are allowed to access the food items freely, but are then immediately injected with a sickness inducing substance so that they form an aversion to the food. So if rats had acquired the goal of accessing the food prior to the aversion, the same goal should now have been devalued and rats should henceforth refrain from pressing the lever. Generally, this prediction holds true. Rats that were subjected to the aversion procedure show significantly lower frequency of lever pressing.

To sum up, if an animal fails in one or the other of the two constraints above, either by retaining acquired behaviour under omission schedules or by performing specific behaviour in the face of goal-devaluation, the behaviour cannot be counted as rational or goal-directed.[57] But how strong is the evidence that animals are truly

57 There is a complication lurking behind the Belief Constraint and the Desire Constraint that I need to mention here. Let us suppose that a scientist proposes a concrete belief constraint and a concrete desire constraint and now goes about testing them in a particular case. Let us further suppose that she finds out that either of the two constraints fails in this particular case. What effect does this finding have for the application of rational behaviour to the animal tested? One answer would be straightforward: The mere fact that the animal fails one of the constraints is enough to show that the animal cannot manifest rational behaviour. But the case is not as straightforward as it first seems. Under certain circumstances, the animal may still manifest rational behaviour even if one of the proposed concrete constraints fails. An example will clarify these circumstances. Allen and Bekoff (1999) claim, for instance, that rats might find it very difficult to change their magazine approach belief under an omission schedule, because the magazine approach behaviour has usually worked very well in the real world. An omission schedule in a laboratory environment, however, is a highly artificial situation. It might thus be more rational for rats to retain a behaviour that has proven so effective in the past if the probability of a particular omission schedule in their natural environment is low anyway. The rat exposed to an omission schedule finds itself with conflicting beliefs. On the one hand a belief, which motivates it not to approach the magazine, and on the other hand a belief, which motivates it to approach the magazine. The rat's earlier experience in natural environments may favour the belief of approaching the magazine, which it then executes. As the explanation shows, the rat behaves completely rational even though the rat does not manifest a change in the investigated belief constraint proposed by the scientist. The source of the complication here lies in the formation of two conflicting instrumental beliefs. The problem for the rat is how to decide when to apply either of the conflicting beliefs. The rat might not be a very good rational reasoner, but we can explain its failings in a rational way – as in fact I just did. These complications notwithstanding, I shall interpret Heyes and Dickinson (1999) as engaging in a project of finding sufficient conditions under which animals manifest rational behaviour. There

capable of engaging in goal-directed action? It is to this question that I next turn.

(A) Evidence for Rational Reasoners

Dickinson and Balleine's (1993) strategy in arguing that rats' actions are properly guided by beliefs and desires highlights a difficulty that a purely associative explanation of animal behaviour cannot solve. Dickinson and Balleine claim to have found a dissociation between properly understood goal-directed actions and mere responses that purely associative accounts cannot explain. Let me outline some of their data.

In a first phase of one of their experiments, Balleine and Dickinson (1991), and Dickinson and Balleine (1993) trained thirsty rats to press a lever for a sugar solution in a single session. After this training session, sickness was induced in the experimental group of rats (EG) by an injection of a mild toxin to condition an aversion to the sucrose solution. The control group of rats (CG) was also made ill but after a delay sufficiently long to prevent any aversion to be conditioned to the sucrose solution. During the second day, both groups of rats were allowed freely to access water in the test chamber in the absence of the lever. And, finally, on the third day, the animals were again exposed to the lever in the test chamber in the absence of rewards. How do the rats perform under these circumstances?

An associative model would predict what would happen as follows. During the training session rats gradually form a connection between the representation of pressing the lever and the representation of the sucrose solution. The actual delivery of the sucrose solution will strengthen the connection between the representations because rats have a positive motivation to access the sucrose solution (since they are thirsty): the rats form a positive goal rep-

might be instances of rational behaviour that their test does not cover, but whatever satisfies their test is thereby a rational behaviour.

resentation of the sucrose solution. As a result of their training, rats will also start to approach the magazine site. The acquired sickness in rats of (EG) will effectively produce an aversion to the sucrose solution because the rats' goal representation has gained a negative qualification. The free exposure to water does not change anything in the above. Consequently, when again presented with the lever, the perception should activate the rats' representation of the lever, and this should in turn activate the representation of the sucrose solution through the connection. Once the representation of the sucrose solution gets activated the rats would understand that the solution is not a desirable substance, and this should finally lead the rats of (EG) to refrain from pressing the lever and to refrain from approaching the magazine site.

Surprisingly, however, this is not what Dickinson and his collaborators observed. Both groups of rats pressed the lever just as frequently. The aversion conditioning did not seem to have left any traces upon the rats' lever pressing and the rats are apparently completely ignorant of its existence. But Dickinson and his collaborators made a surprising observation. While both groups of rats in the experiment showed same performance levels of lever pressing, they nevertheless showed significant differences in their magazine approaches. Rats of (EG) exposed to the aversion conditioning were more reluctant to approach the magazine site than rats of (CG). The aversion is thus clearly manifest in the performance of the approach response of rats of (EG).

Dickinson and his collaborators thus discovered an observable dissociation between the rats' action of pressing a lever and the rats' responses of approaching the magazine. Apparently the rats are ignorant of their acquired aversion to the sucrose solution during training. The occurrence of the dissociation, however, is completely at odds with predictions derived from any associative model (Dickinson and Balleine, 1993).[58] In fact, such dissociation should

58 Dickinson and Balleine (1993) discuss in detail a very powerful associative model in their paper and show that even it cannot account properly for the presence of the dissociation between the rats' actions and responses.

not even occur, since the associative models intend to deliver a unified account of actions as well as responses of animals.

Rats are also capable of learning about their acquired aversion to the sucrose solution as soon as they are given access to the solution. This latter claim has been born out by a slight modification of the Dickinson and Balleine (1993) experiment above. Instead of giving rats of (EG) free access to water after the aversion conditioning in the original experiment, they are given free access to the sucrose solution instead. Accessing the solution gives the rats the possibility of learning about their aversion to the taste of it. As a consequence, rats of (EG) should therefore refrain from pressing the lever once they are exposed to the lever again. This is exactly what Dickinson and his collaborators found: the rats of (EG) refrained significantly from pressing the lever. So the actual experience of the taste of the sucrose solution gives the rats the possibility to make the necessary changes to their goal representation to perform correctly on the test trial.

The results of the experiment indicate where the associative model goes wrong. Note that the model assumes that actions (lever pressing) and responses (magazine approach) are mediated by one and the same goal representation of the sucrose solution. This explains why the model is not able to account for the dissociation. Given that only one and the same goal representation is in play, the aversion conditioning should have changed the representation to accommodate the aversion and consequently should have prevented the rats from pressing the lever. But they still press the lever, even though they respond appropriately by not approaching the magazine. So the data show at the very least that actions (lever pressing) and responses (magazine approach) cannot be mediated by the same goal representation. While one type of "goal representation" may account for the response of approaching the magazine in the associative model, another type of goal representation must mediate the rats' action, which is not accountable in the associative model alone.

Once this difference between the goal representations is postulated, it is easy to account for the observed dissociation between actions and responses along the following lines.

(1) Rats possess goal representations that mediate their goal-directed actions. Let us call them *desires*.
(2) Rats also possess different goal representations that mediate their responses. Let us call them *goal elements*.
(3) *Desires* and *goal elements* do not automatically interact with each other.

Given these conditions, we can now understand why rats show the observed dissociation as long as they are not given access to the solution after acquiring an aversion to it. As the response of approaching the magazine shows, the aversion has its effects, but it is the goal-elements that manifests the avoidance's presence. We can conjecture that the goal elements belong to a basic biological system that determines the motivational reactions to experiences the organism receives. Due to condition (3) the rat's desires underwriting its actions don't automatically interface with this basic biological system. The rat actually needs to expose its desires to the biological system by interacting with the system to learn about its own motivational reactions to the things it desires. Letting the rats access the sucrose solution after aversion conditioning allows the rats to learn about their motivational reactions to them. As a consequence, they can adjust their desires appropriately, and, hence, refrain from pressing the lever. As Dickinson and Balleine (1993, p. 288) conclude:

> Given this characterization of goal 'representations', an interpretation of the outcome of our devaluation study is that intentional goals, or in other words the content of desires, have no direct contact with the processes that actually determine whether a particular stimulus or event is desirable, namely whether the stimulus, either directly or via associative links, activates the reward (or punishment) elements in the reinforcement system. Rather, the intentional goals of actions have to be acquired through self-observation or, more specifically, through acquiring beliefs about the attractive or aversive properties of stimuli and events through observing one's affective reactions to them.

In summary, then, the evidence presented above proves that two completely different types of representational systems, one rational, the other associative, mediate actions and responses in animals.

Because the two representational systems work apart from each other, they need to come into contact with each other so that the animal can learn about its motivational reactions to the various goals it sets itself. But once this happens, animals are capable of showing proper rational or goal-directed behaviour as depicted in the practical syllogism above.

All things considered, Rational Reasoners correspond most closely to Dennett's Popperian creatures, if we take into account his sharp distinctions between Skinnerian creatures with associative principles and Popperian creatures with representational systems. Moreover, if we take the concept of a representational system as fundamental in an evolutionary taxonomy, as we should, we might say that Dennett's Popperian creatures actually include two very different organisms that need to be distinguished from each other to gain an appropriate evolutionary understanding of cognitive capacities: First Representers and Rational Reasoners. Should one still harbour doubts whether First Representers could count as employing explicit representations in their cognition, Rational Reasoners will surely alleviate these doubts. It is hard to make sense of Rational Reasoners' cognition unless one assumes that they have cognitive mechanisms that can access explicit representations, and manipulate them according to the practical syllogism to derive valid anticipations about their environment. The evidence outlined above even shows that Rational Reasoners contain two very different representtational systems that need to interface with each other.

Furthermore, as was already the case with First Representers, we can now introduce considerations already familiar from evolutionary psychology. Evolution's trend the specialisation will have driven species along a path of increasing cognitive diversification. As we have also seen, Rational Reasoners bring a further complication into the picture. Not only must we reckon with cognitive modules that only employ associative principles in the manipulation of representations, but we also have to reckon with more sophisticated cognitive modules that use rational principles to manipulate their representations. The most interesting insight derived from the studies of Dickinson et al. is that these two very different kinds of modules have to interact with each other to let the animal behave in advantageous ways in its environment. What this picture shows us is a very complex system of interactions between cognitive modules. As a consequence of these complex interactions, Rational Reasoners will experience even more selection pressures to solve the modular interaction problems outlined in chapter III. We can then see how, as soon as my proposed conditions of chapter III are met, selection pressures would arise that may be conducive to the development of general-purpose cognitive mechanisms.

3. Conclusion

The chapter started with the observation that weak versions of evolutionary psychology assume that non-human animals possess genuine cognitive capacities that are related taxonomically to human cognitive capacities. I then suggested that evolutionary taxonomies of such animal cognitive capacities are capable of fleshing out the details of this assumption, and that they are also able to produce synoptic insights into the cognitive evolution by outlining the different cognitive options in systematic ways. The chapter first critically discussed Dennett's evolutionary taxonomy in detail in order to assess whether his taxonomy is able to give an appropriate description of animal cognitive capacities. I then showed that his taxonomy does not give an adequate description of animal cognitive capacities because Dennett's taxonomy is threatened by an unsubstantiated and confused understanding of animal representation. According to Dennett, animals are not able to produce explicit representations, which they can access in reasoning, in order to control their behaviour. Rather, animals are able to produce only implicit representations, which are implemented in their designs, and which are thus not amenable to manipulation in reasoning.

Moreover, since Dennett also believes that explicit representations are necessary for thinking, he maintains further that animals cannot be properly described as possessing thoughts.

I argued against Dennett that these negative results do not follow from findings of animal psychology. My analysis of these findings suggested that animals are capable of using explicit representations in their reasoning. Additionally, the analysis distinguished two different kinds of creatures that are able to use explicit representations in their reasoning – First Representers and Rational Reasoners. I then developed an improved evolutionary taxonomy of animal cognitive capacities that both corrects Dennett's confused notion of animal representation and accords First Representers and Rational Reasoners their proper place in the taxonomy.

The result of the chapter therefore is that evolutionary psychology's core assumption about animal cognitive capacities is validated: Animals in fact possess genuine cognitive capacities of various sorts that can be arranged in evolutionary taxonomies. This last result gives additional support to the consistency of evolutionary psychology as a valuable theory of mind. An additional question arises, however, whether animals are similarly able to realise those cognitive states that are most characteristic for human beings: mental states such as belief, desire intention. The following chapter discusses this question further.

V. Do Animals have Beliefs?

The last chapter defended a specific evolutionary taxonomy of animal cognitive capacities. As we saw, these evolutionary taxonomies are grounded in the existence of cognitive continuums that connect animal cognitive capacities with human cognitive capacities via evolution. Now the most dominant human cognitive states are beliefs, desires, and intentions. The question then arises of whether we can assume that animals are also able to realise these specific cognitive states. Thus, are animals capable of belief and other mental states? The evolutionary logic and the cognitive continuum implied by it seem to give evidence to the claim that animals are indeed able to entertain beliefs. This claim, however, is far from uncontroversial.

Opinions about animal belief are strongly divided. On the one hand, sceptics such as McFarland (1989), Davidson (1982, 1984), and Dennett (1996) claim that we do not have, and probably never will have, conclusive evidence for belief states in the case of non-human animals. On the other hand, defenders such as Allen (1992), Searle (1994), and Dupré (1996) claim that the postulation of animal mental states is clearly justified. The outcome of this debate has clear consequences for projects such as the one undertaken with evolutionary taxonomies. Thus an assessment of the arguments against animal mental states is needed to justify these projects further.

This chapter examines an influential argument from philosophy against animal belief that goes back to a series of papers by Donald Davidson. Section 1. analyses Davidson's claim that in the absence of language we are not able to draw the fine-grained distinctions among beliefs that their intentional nature demands. He thinks that without this ability we cannot sensibly assume that animals have mental states. Section 2. then evaluates Davidson's further claim that the having of beliefs implies the mastery of the concept of belief, and that the concept of belief only emerges in the context of radical interpretation. Since animals do not engage in interpretation of the speech of others, they cannot have beliefs in any genuine sense.

The outcome of the assessment of Davidson's argument will show that his conclusions aimed against animal belief are not warranted.

1. Problematic animal psychology

Davidson's (1982, 1984) basic argument is that animals do not have mental states such as beliefs, desires, and intentions because mental states are necessarily bound up with linguistic competence.

Underlying Davidson's reasoning is his account of "radical interpretation" with regard to another person's verbal behaviour (Davidson, 1980, 1984a). In the context of radical interpretation, interpreters of someone else's language are faced with an interdependence of belief ascription and utterance interpretation. To assign particular meanings to a speaker's linguistic utterances requires interpreters to fix the contents of the speaker's beliefs. But at the same time, the interpreters cannot fix the contents of the speaker's beliefs without first identifying the meanings of the speaker's linguistic utterances. According to Davidson, then, the specification of the contents of beliefs and the meanings of linguistic utterances are two aspects of one and the same project. If Davidson is right in his analysis of radical interpretation, and beliefs cannot be determined in the absence of an interpretation of the speaker's utterances, then what about animals that lack linguistic competence altogether? Davidson is quite explicit in his answer, "the chief thesis [...] is that a creature cannot have thoughts unless it is an interpreter of the speech of another" (Davidson, 1984, p. 157)

Davidson's reasoning rests on several assumptions that have to be spelled out in more detail. His argument about the non-existence of animal thoughts rests specifically on two aspects of mental states. Mental states are (1) intentional states and (2) conceptually articulated. I shall briefly outline these two aspects so we can better understand Davidson's attack.

(1) Mental states are distinctive in that they contain representations of objects that may or may not exist, and represent their objects in one way rather than another. Anything that manifests these features is an *intentional* state. Alice, for instance, may believe that Eros visited Psyche every night, even though neither Eros nor Psyche really existed. Moreover, Alice may believe that Eros was a handsome young man without believing that the son of Aphrodite was a handsome young man, even though Eros *is* the son of Aphrodite. Thus intentional states may represent one and the same object in different ways. They create what is known as intensional contexts in which the substitution of co-referential terms may transform a true belief attribution into a false one, as is the case with Alice and the co-referential terms "Eros" and "the son of Aphrodite".

As a consequence of beliefs' intentional nature, if we want to correctly attribute a particular belief to an agent, we must make sure that we can specify not only the objects the beliefs are about, but also how the agent conceives of them. If, on the other hand, something systematically prevents us from drawing the fine-grained distinctions among beliefs that correspond to the different ways an agent may conceive of an object, we would, as a consequence, lose any right we previously had to talk about beliefs in this context in the first place. Davidson's argument, then, is that in the absence of language we cannot draw the fine-grained distinctions among mental states that are necessary for beliefs' intentional nature.

(2) Mental states are also conceptually articulated. This means that an agent properly possesses a particular belief, only if the agent also possesses those concepts the mastery of which are required in order to have that belief. Alice, for instance, can possess the belief that apples are vegetables only if she can correctly master the

concepts of *apples* and *vegetables*. Davidson (1997, 1999) has recently argued that the capacity for thought and the possession of concepts are fundamentally one and the same thing, just seen from different perspectives. Again, he thinks that both thoughts and concepts are confined to creatures that are capable of language.

Moreover, Davidson's claim that thoughts involve concepts is further complicated by his *concept holism*. According to concept holism,[59] the content of a concept is constituted by its functional role within the larger cognitive economy of an agent: its relations with perceptions, further mental events, and behaviour. There are good reasons for supporting concept holism. It is motivated partly by the fact that many theoretical terms seem to be definable only in conjunction with one another, and not with terms outside a theory. An understanding of Newtonian physics, for instance, is only attainable through the learning of inter-definable terms such as *mass*, *force*, *energy*, etc. A novice acquires these terms by using them together in thought processes, especially in problem solving. Furthermore, concept holism seems also to give a plausible analysis of the meaning of logical connectives. For instance, one can specify the meaning of the connective *and* by noting that certain inferences in propositional logic have a special status for the connective in question – namely from *p and q* one can infer *p*, and again from *p* and *q* one can infer *p and q*. Concept holism extends these insights to all concepts, no matter what their particular type is.

Davidson argues that because thoughts involve concepts, thoughts inherit their holism from the latter. He writes,

> We identify thoughts, distinguish between them, describe them for what they are, only as they can be located within a dense network of related beliefs. (Davidson, 1982, p. 475)

So beliefs have to be part of a sufficiently rich network of inferentially connected beliefs, through which particular beliefs acquire

59 For a discussion of concept holism see Block (1986, 1987), Boghossian (1994), Devitt (1996), Field (1977), Fodor and LePore (1992), and Harman (1987).

their determinate content. Put differently, the relations among beliefs are constitutive of the very identity of beliefs, because the content of an individual belief cannot be divorced from what it entails and what it in turn is entailed by.

In summary, both aspects of mental states, i.e. their intentional nature and their conceptual articulation, give rise to problems in the context of non-linguistic animals. The intentionality of mental states raises the question of whether we can specify an appropriate that-clause in a belief sentence, which captures the content of an animal belief state. And the conceptual articulation of thought raises the question of whether animals can possess proper concepts. I shall discuss these problems in the following.

1.1 Problematic animal belief contents

Are we able to specify an appropriate that-clause within one of our belief sentences that precisely captures the content of an animal belief? Davidson introduces his negative answer with a retelling of Norman Malcolm's (1977) dog example:

> Suppose our dog is chasing the neighbor's cat. The latter runs full tilt toward the oak tree, but suddenly swerves at the last moment and disappears up a nearby maple. The dog doesn't see this maneuver and on arriving at the oak tree he rears up on his hind feet, paws at the trunk as if trying to scale it, and barks excitedly into the branches above. We who observe this whole episode from a window say, 'He thinks that the cat went up that oak tree'. (Davidson, 1982, p. 474)

The problem, of course, is whether we are justified in attributing this particular content of a mental state to the dog. Now we already know that Davidson does not think so, because he maintains that we cannot make sense of the fine-grained distinctions among beliefs in the absence of language.

> [...] unless there is behavior that can be interpreted as speech, the evidence will not be adequate to justify the fine distinctions we are used to making in the attribution of thoughts. (Davidson, 1984, p. 164)

But why should we not be able to explain the dog's behaviour by attributing a belief to it? It seems easy enough. In a first attempt, we may explain the dog's barking up the wrong tree by attributing to it the belief:

(1) Fido believes that the cat went up the tree.

But the oak tree also happens to be the oldest tree in sight and the same tree the cat went up last time Fido chased it. But could Fido similarly hold the following beliefs?

(2) Fido believes that the cat went up the oldest tree in sight.
(3) Fido believes that the cat went up the same tree it went up last time.

By the intentional nature of beliefs, we know that only one belief attribution can correctly specify how Fido conceives of the tree. But now it would seem that in the absence of linguistic competence we could not decide which belief attribution is correct. We cannot ask him, overhear one of his utterances, or read some of his written records. In the absence of these methods, all of the belief attributions above apparently explain Fido's barking up the wrong tree equally well, and there seems to be no way to distinguish between them on the basis of Fido's observable behaviour alone. So if we cannot provide additional constraints, the circumstances may lead to serious levels of indeterminacy. As Davidson observes,

> If we persist in attributing desires, beliefs or other attitudes under these conditions, our attributions and consequent explanations of actions will be seriously underdetermined in that many alternative systems of attribution, many alternative explanations, will be equally justified by the available data. (Davidson, 1984, p. 164)

In fact, indeterminacy may become so rampant, so wild, that it starts seriously to undermine the project of attributing mental states to animals. We just cannot fix the content of the mental states in question. Note that Davidson does not merely highlight an epi-stemic point about our ways of coming to know about contents of mental states. He is clearly after a metaphysical point: if we cannot make sense of a mental state having a particular content, we are not speaking about mental states at all. The problem is not so much that we could not specify a content: we clearly can, as shown in (1), (2), and (3). The problem is that we have too many contents and no possibility of singling out the correct one.

Furthermore, the situation is aggravated by Davidson's claim that thoughts are conceptually articulated. Crediting Fido with the belief in (1), for instance, requires that Fido must master the concepts of *cat, tree, etc.* But according to concept holism, Fido can master the concept of *tree* only if he also masters a sufficiently rich network of interrelated concepts such as *leaves, soil, water, burnable materials, growing things, etc.* By the same token, his beliefs about trees will have to be located within a dense network of related beliefs, i.e. beliefs that trees need soil and water to grow, that trees can burn, that trees have leaves or needles, etc. But Davidson thinks that crediting Fido with such concepts would imply too much conceptual sophistication on Fido's part, since

> If we really can intelligibly ascribe single beliefs to the dog, we must be able to imagine how we would decide whether the dog has many other beliefs of the kind necessary for making sense of the first. It seems to me that no matter where we start, we very soon come to beliefs such that we have no idea at all how to tell whether a dog has them, and yet such that, without them, our confident first attribution looks shaky. (Davidson, 1982, p. 475)

Many of those other beliefs, it must be conceded, would be too far-fetched for a dog, and therefore the whole belief structure would lose whatever initial plausibility it had. But are we forced to conclude with Davidson that animals cannot have mental states because they cannot share concepts such as *trees* with us?

A number of philosophers have resisted this conclusion (Allen, 1992; Carruthers, 1992, ch.6; Dupré, 1996; Dennett, 1996, ch. 2; Searle, 1994). John Searle, for instance, remarks that the fact that we cannot specify the contents of animal concepts in our belief sentences does not prove that animals do not possess concepts at all. He notes,

> even if we assume that there is no fact of the matter as to which is the correct translation of the dog's mental representations into our vocabulary; that, by itself does not show that the dog does not have any mental representations, any beliefs and desires, that we are trying to translate. (Searle, 1994, p. 211)

I believe Searle's observation is correct. The fact that we cannot specify in one of our belief sentences how the dog conceives of its environment cannot conclusively prove anything; definitely not the fact that the dog does not possess intentional mental states about its environment. Rather, our inability to specify the dog's mental states should make us suspicious about the sorts of descriptions we use to normally or typically attribute mental states to animals. Notice that in our tree example above, the question was whether Fido had something like the full complement of the concept *tree* that we would expect of another human about whom we asserted "he believes that the cat went up that oak tree". But is this basic assumption appropriate?

Peter Carruthers (Botterill and Carruthers, 1999; Carruthers, 1992, ch. 6) and Daniel Dennett (1996) claim that it is not, because a correct response to Davidson's argument is to insist that the conceptual descriptions Davidson uses to describe Fido are falsely attributed to dogs and animals more generally. Most probably Fido does not have concepts like *tree, soil, water, leaves, etc.* that it can use in its thought processes. What we should do instead is to describe the ways in which the animals themselves represent their environment. As Carruthers (1992, ch. 6) correctly observes, it seems plausible that Fido, being a dog, will represent its environment in terms of some complex set of properties of sight, smell, and audition. So instead of using our concept *tree*, Fido may use something like the concept *scent-markable vertical bar, etc.* Assuming for the moment that such dog-specific conceptual networks are possible, one should refrain from describing Fido's belief as:

(1) Fido believes that *the cat* went up *the tree,*

but should rather render it thus:

(D1) Fido believes that *chasable animate thing like this* (reference to a cat) went up *scent-markable vertical bar here.*

It is to be expected that the ways in which animals represent their environment, given their different ecological niches, will not be easily discoverable by us. That is the reason why we need animal psychology and ethology. Among other things, the sheer difference and alien nature of much of animal mental representation may make it hard to specify animals' modes of representation in our own vocabularies. Moreover, each species may employ its very own modes of representation, which would also further complicate our efforts to arrive at general descriptions. But the complications surrounding animal thoughts cannot be an argument for saying that such modes of thinking do not exist. This seems to be a very weak argument against animal thinking. Davidson's considerations so far are not able to exclude the possibility of animal concepts.

1.2 Circumventing the animal belief content issue

A further criticism of Davidson, and one that Davidson himself anticipates, is to circumvent all of the considerations above. Maybe we can get by without having to specify anything about the animal's conceptual repertoire. A possible response to the difficulties with animal mental states is to hold that in the belief sentence:

(1) Fido believes that the cat went up the tree.

the expression "the tree" occurs in a transparent position of the sentence, i.e. it occurs in a position that allows substitution of co-referential terms without changing the truth-value of the sentences. We could paraphrase (1) to make the transparent position of "the tree" more explicit as:

(1*) Fido believes *of the tree* that the cat went up it.

Now substituting "of the oldest tree in sight" and "of the tree the cat went up last time he chased it" for "the tree" won't change the truth-value from one belief-sentence to the next. So if it is true that

(1*) Fido believes *of the tree* that the cat went up it.

it still remains true, for instance, that

(2*) Fido believes *of the oldest tree in sight* that the cat went up it.

While the "believe of" locution still has to specify a referent from within the conceptual framework of the attributer, it relieves the attributer from specifying anything specific about the conceptual repertoire of the believer. Accordingly, we may hope that our use of belief-sentences such as (1) proves enough to give us an explanatorily satisfying account of animal behaviour, even though such sentences do not require a specification from the believer's point of view. But it is doubtful whether this reply to Davidson's point is successful. As Davidson observes,

> But such constructions, while they may relieve the attributer of the need to produce a description of the object that the believer would accept, nevertheless imply that there is some such description; the *de re* description picks out an object the believer could somehow pick out. In a popular if misleading idiom, the dog must believe, under some description of the tree, that the cat went up that tree. But what kind of description would suit the dog? (Davidson, 1985, p. 475)

Davidson's point is apposite. A closer look at a belief-sentence like:

(1*) Fido believes *of the tree* that the cat went up *it*.

shows that "it" requires an anaphoric referent, whose referent Fido has to discriminate from other things in its environment. In order to do this, Fido will presumably employ some mental descriptions that allow him to pick out one referent rather than another. But once we concede Davidson's concept holism and its tenet that particular concepts only make sense within a dense network of further concepts, we see that we run into difficulties. If we are forced to produce a mental description of the object that would suit the animal, and concepts make sense only within a dense network of related concepts, then Davidson's point that belief sentences imply too much conceptual sophistication on the parts of animals may still be valid. While it is true that Davidson's argument originally applied only to human frameworks, his points about conceptual sophistication are now seen to apply to animal conceptual frameworks as well. For even granted that animals possess their own specific conceptual frameworks, concept holism invariably implies that animals must possess rich networks of inferentially connected concepts. As Davidson writes,

> There is no fixed list of things someone with the concept of a tree must believe, but without many general beliefs there would be no reason to identify a belief as a belief about a tree, much less an oak tree. (Davidson, 1982, p. 475)

Substitute for "tree" whatever concepts best suit the animals' conceptual repertoire and the same reasoning about conceptual sophistication may still carry through. But can we reasonably assume that animals demonstrate such levels of conceptual sophistication?

Evidently whether we are justified or not in assuming that animals use concepts depends on the strength of the holism that Davidson employs in his argument. There may be stronger or weaker versions of concept holism that affect the argument in different ways. Davidson's argument against animal belief, however, will only work if he can establish a concept holism that is, first, reasonable enough and, second, strong enough to rule out animals as proper believers. I shall now discuss different versions of concept holism with a view to seeing whether one version of it is effective in ruling out animals as proper believers.

1.3 Different versions of Concept Holism

(A) Strong Concept Holism:

According to strong concept holism, an individual belief's identity is constituted by all the inferential links the belief has to all other beliefs in the network, and an agent possesses the individual belief only if the agent is inclined to make all the inferential links corresponding to the individual belief in question.

Now this condition seems excessively restrictive. For one can easily imagine situations in which an agent believes the axioms of some theory without thereby believing all the consequences that can be inferred from them. Asked about a specific theorem, the agent can truthfully reply that he does not believe it and still maintain that he does nevertheless believe in the axioms of the theory. Initially, at least, we would not be inclined to say that such a person is conceptually confused. Moreover, to suppose that a particular belief's identity is constituted by the entire system of its inferential links makes it highly dubious how two agents may ever share an individual belief, for there will always be enough difference between the two belief systems to generate some differences between agents in the inferential links of the individual belief in question (Fodor and LePore, 1992). But any difference in the inferential links of two beliefs will imply that they are different beliefs altogether, and so no beliefs could be shared by two people. Strictly interpreted, strong concept holism leads into absurdly restrictive positions of belief attribution and should therefore be disregarded as a reasonable version of concept holism.

A much more plausible account of concept holism allows an agent to possess an individual belief without thereby requiring of the agent that she must actually believe every other belief inferentially related to it. A natural reply to strong concept holism's problems may be to weaken the concept holism in some ways. One such way is to no longer require that the agent must *actually* believe every other belief inferentially connected with a particular belief in order to possess it. It would then be seen as sufficient that she only *potentially* believes the other thoughts within the inferential network. This weaker form of concept holism is similar in spirit to a distinction Dennett (1978) uses between *occurrent thoughts* that an agent actually believes at a particular time and all the other *potential thoughts* that the agent may *generate* when asked to or that it is capable of entertaining when appropriately triggered on the basis of its occurrent beliefs. I may, for instance, have the occurrent belief that cheese is yellow without thereby actually believing that the production of cheese is dependent on bacteria. Nevertheless, I am capable of entertaining or understanding the latter belief when asked about it. So two agents may share an individual belief as long as they share the same occurrent beliefs. But this version of concept holism is susceptible to much the same arguments as its stronger cousin. It is true that of the *potential* beliefs the agents are not required *actually* to have all of them. But putting the issue in this way does not make it clear that on another level they are required to assent to the same beliefs surrounding an individual belief – when suitably triggered by questions about them. For if one of the agents is capable of entertaining one belief that is inferentially linked to the target belief whereas the other is not, this difference will manifest that they did not share the specific belief after all. Changing from actually believing something to potentially believing it has not made us any headway so far.

It is true that Davidson does not spell out the strength of his preferred holism in any detail. But passages such as these,

> There is no fixed list of things someone with the concept of a tree must believe, but without many general beliefs there would be no reason to identify a belief as a belief about a tree, much less an oak tree. (Davidson, 1982, p. 475)

> There are good reasons for not insisting on any particular list of beliefs that are needed if a creature is to wonder whether a gun is loaded. Nevertheless, it is necessary that there be endless interlocked beliefs. (Davidson, 1984, p. 157)

show that Davidson should not be read as holding a version of strong concept holism with respect to beliefs. The implication of these passages is, apparently, that a creature can possess a particular belief, if it also possesses *some or many* of the inferential links surrounding the individual belief, but by no means *all* of them. The requirements that there is no fixed list of things an agent has to believe, or the insistence that there is no particular list of beliefs an agent must possess allow this reading of these passages. Moreover, these passages show that two agents may share a particular belief even though they differ in their respective belief networks, as long as their respective belief networks show enough overlapping beliefs. Taking Davidson's suggestions as a clue we can outline a more promising version of concept holism. A much more plausible version of holism, and one that Davidson himself apparently favours, will regard only some proper subset of the totality of inferential links surrounding an individual belief as constitutive of the individual belief's identity. Let us call such a position weak concept holism.

(B) Weak Concept Holism:

According to weak concept holism, an individual belief's identity is constituted by *some* subset of the inferential links surrounding the individual belief, and an agent possesses the individual belief in question only if the agent also believes in the subset of the inferential links corresponding to the individual belief.[60]

Some philosophers have traditionally opted to single out the privileged subset of inferential links by using some version of the analytic/synthetic distinction.[61] Others have attempted to single out the privileged inferences by other means, in terms of "primitively compelling" inferences that surround a given thought or concept (Peacocke, 1992), in terms of those inferential connections which are necessary and sufficient for the truth of some law-like generalisations (Botterill and Carruthers, 1999), or in terms of "centrally explanatory uses" of concepts (Horwich, 1998). The controversy over how precisely to delimit the privileged subclass of inferential links is still unresolved, and is the topic of much on-going debates.

But I need a clear handle on the issue to evaluate Davidson's argument against animal belief. I shall use Peacocke's primitively compelling inferences in the following for the purposes of my argument. Peacocke's account of primitively compelling inferences will help to conceive which inferential links are likely to fall within the privileged class and thus are constitutive for a belief's identity. It is thus necessary to present briefly Peacocke's definition of primitively compelling inferences to get a better idea of his proposal.

According to Peacocke (1992), to possess a concept an agent must find certain inferences primitively compelling in such a way that (a) she would find such inferential transitions self-evident, (b) she would not find them compelling because they were inferred

60 It is true that Davidson himself does not argue for weak concept holism in quite those terms. But I must draw clear distinctions between the different forms of concept holism to evaluate Davidson's claims. As the cited passages make abundantly clear, Davidson wishes to leave these issues open. He does not want to be commited to any doctrine of concept holism. Whatever works, he seems happy to take on board. But for the purposes of my argument a decision about some form of concept holism must be taken, otherwise an evaluation will not be possible.

61 See Boghossian (1994) for a discussion of the analytic synthetic discussion in the context of concept holism. Fodor and LePore (1992) have given a forceful criticism of this distinction.

from other premises, and (c) she need not take the correctness of the transitions to be answerable to anything else. Peacocke (1992, pp. 137-138) analyses, for instance, the mathematical concept *Plus* as that concept C to possess which an agent must find primitively compelling instances of transition (1) and principle (2), and the agent must moreover find them primitively compelling because they are of that form, and do not depend on anything else – m, k, n are variables for natural numbers, s( ) designates the successor of a natural number, and C stands for the concept.

(1) from (mCk) is n
    to (mCs(k)) is s(n)

(2) (mC0) is m

Here is not the place to go into the details of Peacocke's account. I introduce the bare essentials only for the purposes of giving a rough idea of how a proper subset of the totality of the inferential connections surrounding a given concept may be singled out. Taking Peacocke's suggestion into consideration, we can now say more precisely that weak concept holism allows two agents to share a particular belief, as long as they both find primitively compelling the same inferences surrounding their particular belief.[62]

Having introduced these different versions of concept holism, I shall now consider whether Davidson is justified in ruling out animal belief on the basis of strong or weak concept holism. Put simply, I do not think he is justified in ruling out animal belief on either version of concept holism. The reasons underlying my claim are straightforward. We can put aside strong concept holism from the start. As we have seen, Davidson apparently does not want to

62 As may be the case, the condition depends on both agents being able to draw the same sorts of 'primitively compelling' inferences. Given the fact that members of the same species will share similar cognitive structures, the condition above appears to be well founded. For a related problem and discussion of it see Sperber and Wilson (1986).

commit himself to strong concept holism anyway. Once strong concept holism is eliminated, the only position left is weak concept holism. But even weak concept holism will not give Davidson an argument against animal belief. To see this, let us reconsider Fido's belief:

(1) Fido believes that the cat went up the tree.

According to weak holism, two agents share a belief as long as they share some privileged subclass of inferentially connected beliefs. It can be proven that Davidson's conclusion against animal beliefs cannot follow from weak holism alone by using weak holism first in conjunction with the assumption that animals can share concepts with us and second in conjunction with the assumption that animals possess their own particular conceptual repertoire. On both assumptions the conclusion is not derivable from weak concept holism alone.

Criticism 1: Assuming animals can share concepts like ours

Let us assume that the question is whether Fido is capable of sharing beliefs about *trees* with us. Now we already know that Davidson claims that Fido cannot, because our beliefs about *trees* are embedded in rich inferential networks of further beliefs such as beliefs that *trees burn*, or that *trees* need *water*: beliefs which are beyond Fido's grasp but without which we cannot make sense of Fido believing anything about trees. Now weak concept holism requires that two agents can share beliefs as long as they share the primitively compelling inferences surrounding a particular belief. To justify his conclusion against animal beliefs, Davidson would have to show that his proposed list of beliefs must be included within the primitively compelling inferences surrounding the belief that the cat went up the tree, *and* that Fido cannot reasonably possess them.[63]

63 As already indicated in footnote 2 above, Davidson does not put the problems discussed here in terms of primitively compelling inferences. I am

There is no good reason to think, however, that Davidson could establish such an argument to the conclusion that his proposed list of beliefs belong to the primitively compelling inferences. Whatever the privileged subclass turns out to be, it seems *prima facie* possible that one can believe something about *trees* without believing that *trees burn* or that *trees* need *soil* and *water* to *grow*. Short of arguing for strong concept holism, it is not easy to see why such beliefs should be included in the privileged subclass of inferential links that constitute a belief about trees. Surely they do not belong to the primitively compelling inferences an agent must make to possess the concept *tree*. The inferential transitions from:

(1) Fido believes that the cat went up the tree.

to

(1') Fido believes that the cat went up something that is burnable.

or

(1") Fido believes that the cat went up something that needs water and soil to grow.

do not appear to be self-evident in the required sense. There is a lot of empirical lore about trees underlying those transitions. Additionally, depending on such empirical generalisations about trees, the inferential transitions are themselves inferred from other premises, namely myriad empirical observations, and their correctness is also answerable to such observations. So the inferential transitions do not satisfy Peacock's definition of primitively compelling inferences. Presumably, *some* of the beliefs included within the primitively compelling inferences will be based on the direct perceptual

helping myself to Peacocke's proposal as a tool to understand and evaluate the strength of Davidson's argument against animal beliefs. I thus cannot claim that Davidson is committed to anything like Peacocke's proposal, but as long as Davidson does not give arguments for his preferred version of concept holism, I am taking Peacocke's proposal as a reasonable elaboration of a form of weak concept holism.

appearance of trees, i.e. beliefs that *trees* have *branches*, or that they are *green*, that they grow *vertically*, etc. But nothing so far has shown that Fido could not in principle hold *such* perceptual beliefs. Clearly more has to be done to exclude animals from having such beliefs.

But what about beliefs that are somehow more strongly connected to concepts of natural kinds such as, for instance, the concept of cat: namely beliefs that *cats* are *animals*, or that *cats* are *enduring physical objects*, etc.? Could someone truly believe that *cats* climb *trees* unless she also believed that *cats* are *animals*, or that *trees* are *inanimate things*? And could she believe in the latter without believing that *cats* and *trees* are *enduring physical objects*? Well it is harder to see how someone could believe anything about cats without inferring that cats are animals, and so on. For a start, the inferential transitions from cat to animal, or tree to inanimate thing appear to be primitively compelling, etc. These examples cannot be as easily dismissed as the other cases above. But if those beliefs belong to the privileged subclass of inferences, we will be forced to attribute concepts of *animals*, *inanimate things*, and *enduring physical objects* to animals. But is that not too much conceptual sophistication for an animal? If Davidson were correct in claiming that such beliefs are beyond the capacities of animals, weak concept holism would indeed be enough to rule out animals as proper believers.

The question is of an empirical nature. Only empirical evidence can tell whether animals are capable or not of entertaining these sorts of concepts. But we may nevertheless try to make some conjectures about the structure of animal concepts such as those above. I shall venture such a conjecture that benefits from the considerations established in earlier chapters to show that weak concept holism may not, after all, be strong enough to rule out animals as proper believers.

As our discussion of the logic of the evolutionary process in chapter I and the evolutionary taxonomy introduced in chapter IV showed, animals might use multiple cognitive modules to face their environments. The complexity of the knowledge embedded in those modules is of such a nature as to suggest that it is conceptually, rather than just perceptually, based (Botterill and Carruthers, 1999, ch. 3; Sperber, 1994).[64] Granted that this picture of cognition is not completely on the wrong track, we could thus conjecture that some of the creature's conceptual repertoire is innately pre-configured by cognitive modules and the particular conceptual resources they make available (Botterill and Carruthers, 1999, ch. 5). Accordingly, minds could be innately partitioned by concepts such as *animal, inanimate thing, enduring physical object, people,* or whichever concepts are provided by the activity of specialised cognitive modules.

Each module, in turn, is seen as a pre-configured and hierarchically organised knowledge structure that correctly situates incoming input. Put differently, information that satisfies the correct input-conditions of modules will be associated with the sorts of inferential links that are specific to the conceptual elements and their relation to each other, which is provided by the module at hand.

To capture the primitively compelling character of the transitions from, for instance, *cat* to *animal* or *enduring physical object,* the current conjecture postulates two very different cognitive modules, one module concerned with problems of animate/inanimate behaviour and the other concerned with the behaviour of physical objects generally. While these modules work in relative isolation from each other, we can nevertheless imagine that they somehow interface with each other. Based on this picture of cognition we can understand how Fido may be compelled to draw the following inferential transitions:

from

(1) Fido believes that the *cat* went up the *tree.*

to

(3) Fido believes that *cats* are *animate.*

64 This means that cognitive modules do not only receive perceptual input that is then suitably conceptualised and fed into higher conceptual abilities. Cognitive modules may also receive suitably conceptualised input from further down the line which they further compute before the output is further processed, and so on. See Chapter I for more details.

(4) Fido believes that *trees* are *inanimate.*
(5) Fido believes that both *things* are *enduring physical objects.*

The inferential transitions follow closely the informational flow through Fido's cognitive architecture. Whenever Fido perceives a typical perceptual outline of a cat, his *cat* concept is activated. Once that concept is activated it is fed into the cognitive module concerned with the distinction *animate/inanimate.* This latter module associates the cat concept with the relevant inferential connections typical of its specialised knowledge structure, so Fido will be primed to thinking that *cat* IS *animate.* As a result of the modular resources, the *cat* concept is enriched by the specialised knowledge of the module, i.e. Fido will also be primed to think that *cat* IS capable of *self-propelled movements,* and all the other features that are characteristic of animate behaviour. Moreover, if we assume that the current cognitive module interfaces with the cognitive module concerned with physical objects generally, we can see how, once the object module is fed with the concept *animate,* it will yet again associate the right inferential connections with the incoming concept. Fido will now be primed to think that *animate* IS *physical object,* and the concept *animate* will be further enriched by the specialised knowledge within the module, e.g. *animate* IS *independent of one-self, exists even if not perceived, etc.*

According to this picture of cognition, Fido cannot possibly hold the belief that *Felix* is a *cat,* unless he is also ready to infer that Felix is *animate,* and that Felix is an *enduring physical object.* It need not be the case, though, that Fido has to consciously or reflectively draw these compelling inferences. It may rather be the case that his cognitive modules draw the inferences for him. Nonetheless, one sees how one can give a natural explanation of primitively compelling inferences on the basis of cognitive modules. New results in animal psychology show that the concepts used above may be available to animals (Tomasello and Call, 1997; Russon, Bard, and Taylor Parker, 1996; Byrne and Whiten, 1988). I cannot adduce all the relevant evidence here, so I shall briefly outline only two instances.

One instance is that primates, in general, appear to understand essential properties of *physical objects* such as object continuation. Chimpanzees, for instance, readily pass the A-not-B test introduced by Jean Piaget. In the A-not-B test, subjects first find an object under a cover A, and later watch as an experimenter moves the object from under cover A to under another cover B. Human infants up to about 12 months of age are not capable of solving this problem because they continue to search where they last found the object, namely under A even though it is now under B. After 12 months of age, infants are able of tracking the visible displacement of the object from A to B and to solve the problem by looking under cover B. The solving of the problem implies a knowledge that objects can move to new locations independently of one's own actions upon them. As has been repeatedly shown, primates are capable of solving the problem and thus demonstrate a basic understanding of the continuance of physical objects (Tomasello and Call, 1999, ch. 1); they thus can apparently master the concept of *enduring physical object*.

Another interesting set of data is that primates, in general, appear to understand something about the distinction between *animate* and *inanimate* behaviour. In an experiment conducted by Marc Hauser (1998) with capuchin monkeys, the monkeys are exposed to two identical chambers that have clear Plexiglas front panels separated by an opaque partition. In the test trials, capuchin monkeys watched as an object was placed in one of the chambers. Both chambers were then momentarily occluded from the subjects' view, and later the occluder was removed again revealing the test object in either the same chamber or the opposite chamber. If the object appears in the opposite chamber, monkeys should infer that the object has moved based on its own by self-propelled motion. Now depending on whether or not monkeys have a specialised knowledge about animate vs. inanimate behaviour, they should show surprise at finding an *inanimate* object in the opposite chamber, whereas they shouldn't show surprise at finding an *animate* object in the opposite chamber (since its animate behaviour is self-propelled).

Hauser (1998) measured the surprise of monkeys by registering their preferential looking times toward one or the other chamber, the assumption being that animals show more surprise whenever some of their expectations are violated in particular situations. The result was that subject failed to show a statistically significant change in looking times to a live mouse, regardless of whether they saw the mouse on the same or the different side from where it was originally placed. However, looking times increased significantly when some Fruit Loops cereals appeared in the opposite side. These are then clear results that monkeys possess knowledge about the distinction between animate and inanimate behaviour. Moreover, Hauser (1998) was able to differentiate the monkey's reactions to self-propelled, but inanimate objects (a clay face that is moved by a hidden magnet) and self-propelled, but animate objects (mouse, frog). The results appear to show that monkeys are specifically attuned to understand animate behaviour. They thus appear to be capable of using the distinction *animate/inanimate* in their reasoning about their environment.

Instances like these suggest that animals do possess conceptual abilities that go far beyond the direct perceptual information given to their senses. Incidentally, the modular position introduced above can also deal in a similar way with criticisms raised by John O'Leary-Hawthorne (1993) in support of Davidson. Before showing how the modular position can do this, I shall very briefly introduce O'Leary-Hawthorne's points, which should in any case already be somewhat familiar through my discussion of the network constraint of belief.

O'Leary-Hawthorne raises problems for theories that want to attribute (relatively) simple beliefs about the current environment to animals. O'Leary-Hawthorne claims that "the content of one's be-liefs about the environment is not merely a matter of the sorts of experiential situations in which they occur but also of their role in one's theory of the world." (1993, p. 467) It is clear that O'Leary-Hawthorne intends to spell out the details of the network constraint of belief. So he basically points out two issues that surround beliefs: issues he thinks that undermine our attribution of mental states to animals.

(a) Quine (1960) has shown that beliefs about the current environment are frequently affected by "collateral information" and are thus open to the influence of surrounding beliefs. So, for instance, "whether we take a feature of the environment as supporting a causal belief of the form 'A will cause B' will depend a lot on our background beliefs about causation – about what sorts of features are causally efficacious, what causal generalizations are true, and so on." (O'Leary-Hawthorne, 1993, p. 479) So even simple beliefs about particular features of the environment are open to the influence of theory-like beliefs.

(b) Additionally, theory-like beliefs are dependent on data they need to explain. However, in the case of non-linguistic creatures a major problem arises.

> When we have a language, it is pretty clear what data our theory needs to explain. That is because we have records of what experimental results have been previously observed that are written down or typed into a computer. [...] If a languageless creature is to employ a theory, however, its assessment of the data that the theory needs to explain will be a function solely of what it remembers about its previous encounters with the environment, together with beliefs about what it is currently observing. (O'Leary-Hawthorne, 1993, p. 471)

As a result the theory-like beliefs we must attribute to an animal will depend upon what we take its beliefs about the past to be, and without language we will not be able to pin these down precisely.

Both of O'Leary-Hawthorne's considerations are intended to show that, in the absence of language, indeterminacy becomes so wild that it effectively undercuts the postulation of mental states in the case of animals. Now, the modular position I suggest above may show why O'Leary-Hawthorne's points are not really an improvement over those of Davidson. For a start, modules help to narrow down considerably the range of possible interpretations of a problem. Cognitive modules help the organism to interpret particular information in the light of specialised knowledge and such modules may thus take care of the effects of collateral information. So, for instance, whether we take a specific generalisation in the environment as causal or not will depend a lot on our background beliefs about causation. It will depend on our folk-theory about causality – O'Leary-Hawthorne is right so far. But such a theory may be realised in specialised modules that help the organism to single out causal regularities from non-causal regularities in its environment. As we saw in Hauser (1998), animals are capable of singling out certain features of the environment (animate behaviour) by applying an innately specified "theory" of animate/inanimate behaviour. A similar explanation that postulates a specialised module about the causal behaviour of objects could equally explain how animals can draw on data about causality beyond "what it [the animal] remembers about its previous encounters with the environment, together with beliefs about what is currently observing" (O'Leary-Hawthorne, 1993, p. 471). Modules embody the recurrent features and patterns of situations, which ancestors repeatedly encountered in their evolutionary environment. By having, for instance, a module dealing with causation, the animal can take advantage of past data that contributed to the development of the module. So the animal is not strictly committed to the data it is currently observing.

Of course, animal psychologists need to single out precisely what and how many modules animals possess. But modules are an effective way to deal with the points raised by O'Leary-Hawthorne. In so far as O'Leary-Hawthorne disregards the modular option as a solution to his worries, his critical points cannot succeed in ruling out animal mental states. With this last failure to argue against animal beliefs, we can now maintain that weak concept holism is too weak to be used to argue against animal belief since it does not include anything that excludes animals from satisfying the constraints imposed by it.

Criticism 2: Assuming animals possess their own concepts

Another way of seeing that weak concept holism will not buy Davidson his preferred conclusion against animal belief is as follows. As Carruthers (1992) and Dennett (1996) observe, it may be completely inaccurate to attribute human concepts to animals in the first place. Humans conceive of their world with the aid of a language capable of drawing fine-grained distinctions among beliefs. Moreover, humans can use the resolving powers of their languages in their communication with each other to express their particular human modes of representing their world. But for animals that cannot conceive of their world with the aid of languages, and which rather rely on their own particular conceptual repertoires, we cannot directly use language to *express* the animal's particular modes of representing their world (Dennett, 1996). Language's resolving power just cuts too fine for animal conceptual systems. But we can nevertheless use language to exhaustively *describe* their ways of representing their world (even though we cannot express them). If we can figure out and describe how animals conceive of their world, and how their conceptions work together (and weak concept holism may give us an inkling here), we will know everything there is to know about their particular conceptual repertoires. Simply showing that animals cannot possibly possess concepts that linguistic creatures do does not show that animals cannot possess concepts at all: it just shows that they do not possess the *same sorts* of concepts.

The most Davidson's argument is able to show is that if we want to attribute conceptual abilities to animals, we had better make sure that these beliefs are embedded within networks of inferentially connected beliefs that satisfy weak concept holism. And nothing so far has demonstrated that animals' beliefs cannot do that. In fact, a modular view of cognition has ample resources to suggest that animals can draw the required inferences through pre-specified conceptual repertoires. So whichever way we turn, Davidson's conclusion against animal belief is not justified.

2. The concept of belief

But there are additional worries about animal thought that need to be addressed. Recall that when I first introduced sentences of the form:

(1*) Fido believes *of the tree* that the cat went up it.

I said that the "it" forces one to specify the ways the animal conceives of the objects it represents. I assumed in the above discussion that animals are capable of conceiving objects in a particular way on the basis of their own conceptual repertoires. Once this was granted, the arguments proceeded to show that Davidson could not rule out the existence of animal beliefs.

However, the very assumption that animals use full-fledged concepts to discriminate among objects in their environment may be doubted. In fact, Davidson doubts precisely this assumption when he states that,

> A creature may react to the world in complex ways without entertaining any propositions. It may discriminate among colors, tastes, sounds and shapes. It may [...] change its behavior in ways that preserve its life or increase its food intake. It may 'generalize', in the sense of reacting to new stimuli as it has come to react to similar stimuli. Yet none of this, no matter how successful by my standards, shows that the creature commands the subjective-objective contrast, as required by belief [beliefs that involve concepts]. (Davidson, 1985, p. 480)

Beneath Davidson's observations lies a distinction he makes between *classifications* based on concepts and mere *discriminations* based on discriminatory abilities. There is undoubtedly an important difference between cases in which animals are merely responding to features in the environment without forming mental states in which *concepts* figure, and cases in which animals are responsive to those features of the environment because they form *concepts* of them. For example, animals might be conditioned to respond in a certain way to a particular tone, yet we might be less than certain in crediting them with the concept of *tone*.

Evidently the argument depends on a precise understanding of discrimination in contrast to classification. As it turns out, drawing a precise difference between these two activities is particularly hard. But I believe it is fair to say that Davidson thinks of discriminations as purely perceptually based dispositions that, in one way or another, automatically produce behavioural responses in the presence of well-defined perceived situations. They are the stimulus-response connections once favoured by behaviourism. Whether or not these dispositions use representations mediating the behavioural responses,[65] those representations if they do exist, will be tied to the information given to direct perception, and will not in any way go beyond such information. The only way discriminations are capable of transcending the actually given information is, as Davidson remarks, by generalising their responses to similar stimuli situations.

By contrast, classifications involve a *normative* dimension. To classify things into those, which are F and those, which are non-F, it must be possible that an agent should sometimes be mistaken and classify Fs as non-Fs and non-Fs as Fs. Thus classification involves an ability on the part of the agent to recognise her own mistakes. Davidson refers to this ability in the passage cited above when he writes that animals do not command the subjective-objective contrast. To have a concept, the agent must recognise that sometimes she may be *subjectively* mistaken and classify an object as something, which in fact it, *objectively*, is not. According to Davidson's claim that concepts and the capacity for thought are closely tied to each other, beliefs as mental states will inherit the normative dimension from the concepts they contain. So an agent will be capable of entertaining beliefs only if she can understand the possibility of being mistaken. Davidson (1975, 1985) has mounted an influential

65 See chapter IV for a detailed discussion about these different interpretations of stimulus-response connections.

argument on the basis of these last considerations that may be summarised briefly as follows:

(P1): If agent A has a belief, then A possesses the concept of *belief.*

(P2): If agent A has the concept of *belief,* then A possesses the concept of *objective truth.*

(P3): If agent A possesses the concept of *objective truth*, then A must be an interpreter of someone else's speech.

(C): Therefore, if agent A has a belief, A must be an interpreter of someone else's speech.

We can see how the conclusion excludes animals from believers. The inference is logically valid, but its soundness depends on whether or not the premises are true. An attack on Davidson's argument must therefore show that one of its premises is not true. In the following I will criticise Davidson's arguments on the basis of the truth of its premises. But before engaging with the argument above I need to flesh out the reasoning behind Davidson's premises.

In "Rational Animals" (1985), Davidson maintains that an agent who believes something thereby generates expectations about how the world is. But expectations can go wrong sometimes in the case when the world does not turn out to be as we expected it. It is characteristic for expectations that they carry with them an awareness of possible frustration. This awareness is expressed in the agent's *surprise* when the world is not what she thought it to be. For Davidson, *surprise* shows an awareness of a contrast between what the agent once believed and what it now comes to believe on the basis of new facts. So what it comes down to, according to Davidson, is that showing surprise is correlated with commanding, however tacitly, the subjective-objective contrast. Moreover, an agent who is surprised forms a belief that her original belief was false: she thereby forms second-order beliefs about first order beliefs to the effect that they are wrong. Put differently, we can say that the agent possesses a concept of *belief* in virtue of which the agent can conceptualise her beliefs *as beliefs*. Evidence of the important theoretical role surprise plays for Davidson is found in the following passage:

> Surprise requires that I be aware of a contrast between what I did believe and what I come to believe. Such awareness, however, is a belief about a belief: if I am surprised, then among other things I come to believe that my original belief was false. I do not need to insist that every case of surprise involves a belief that a prior belief was false (though I am inclined to think so). What I do want to claim is that one cannot have a general stock of beliefs of the sort necessary for having any beliefs at all without being subject to surprises that involve beliefs about the correctness of one's own beliefs. (Davidson 1985, p. 479)

Accordingly, Davidson believes that these considerations can discharge (P1) in the argument above: to have a belief, an agent must master the concept of *belief*.

To master the concept of *belief*, and thus to know what beliefs *as beliefs* are, is to grasp the contrast between objective truth and mere subjective opinion: true belief and false belief. According to Davidson, then, mastery of the concept of *belief* presupposes the concept of *objective truth*. This strategic move allows Davidson to discharge (P2) in the above argument.

In a final move, Davidson attempts to demonstrate that the concept of *objective truth* is anchored in interpretation where one agent tries to make sense of the speech of another. Some explanations are in order here to indicate what work the concept of objective truth does in interpretation.

Recall that Davidson maintains that the project of radical interpretation is made more difficult by the interdependence of belief ascriptions and utterance interpretations (Davidson, 1984). I can only know what a speaker means by one of his utterances if I know what he intends the utterance to express, and I can get access to the speaker's beliefs only by first identifying the meanings of his linguistic utterances. Davidson suggests that an interpreter can break into this circle by holding constant the speaker's attitude of holding true a sentence under particular circumstances. It is possible for a radical interpreter to do this because he can detect the speaker's attitude of holding true a sentence without previously knowing what the sentence means.

The first and basic data of a theory of radical interpretation are sentences that are strongly correlated with observable circumstances. Sentences of the form:

Speaker s holds true the utterance: "It rains" *at t and it rains at t in the vicinity of s.*

Based on these observation sentences, the radical interpreter assigns meanings to them and thus starts slowly breaking into the speaker's belief-meaning circle. However, the radical interpreter will be able to develop an accurate interpretation of the utterances of his speaker only if he can assume that the sentences the speaker holds true really are true in the observable circumstances. Clearly this will not always be the case, since his speaker will be mistaken some of the time. According to Davidson, though, a fundamental condition of successful interpretation is that we can dismiss a priori the possibility of massive error. It must therefore generally be the case that a sentence is *objectively true* when a speaker *subjectively* holds it to be true – otherwise successful interpretation would not be possible.

In the contrast of sentences, which are *objectively* true and sentences, which are only *subjectively* held to be true, the project of radical interpretation finally assigns the concept of *objective truth* a systematic place. Davidson would maintain that we have the idea of objective truth only from the role it plays in the interpretation of a language (Davidson, 1984). With these last considerations, Davidson can finally discharge (P3) of his argument above and reach his preferred conclusion.

As I already indicated, one must object to Davidson's argument by objecting to one of its premises. The premise on which I want to concentrate here is (P1): that to have a belief one needs to have the concept of belief. As I showed before, Davidson justifies (P1) by asserting that having beliefs requires the possibility of being surprised. Such surprise, Davidson claims, is a belief about a belief. Accordingly, if someone is surprised, then she comes to believe that one of her original beliefs was false, and she should revise her believe to take account of that fact.

The first part of the argument seems reasonable enough. Having a belief requires some cognitive mechanisms that can:

(a) monitor whether a particular belief has been satisfied or not
(b) initiate some sort of self-corrective behaviour that results in the revision of one's false beliefs

Now Davidson thinks that these cognitive mechanisms are conceptual or reflective activities, i.e. second-order thoughts that tell us what we ought to believe. Moreover, they let us grasp the contrast between true belief and false belief. But is Davidson justified in assuming that the cognitive mechanisms are conceptual in this way?

Davidson apparently fails to notice an important alternative: namely the possibility that the animal grasps the contrast between true belief and false belief implicitly through its self-corrective behaviour (Carruthers, 1992, ch. 6; Searle, 1994). We may say that an animal is aware of the possibility of mistakes if it is somehow capable of revising its beliefs in the light of contradicting evidence. But this belief revision need not involve consciously reflecting upon the beliefs in question. Belief revision may be driven instead by implicit procedures that are capable of assessing the success or failure of particular mental states. Animals understand when their occurrent desires are thwarted by circumstances, so they may also understand when some of their beliefs are not satisfied by how the world in fact is. As Searle (1994, p. 212) remarks, "true" and "false" may function as *meta-intentional markers* that "are used to assess success and failure of representations to achieve fit in the mind-to-world direction of fit [...]"

Rats, for example, may be able to demonstrate the required abilities of self-correction. As we saw in chapter IV, when rats are repeatedly exposed to the contingency of pressing a lever followed by food, they easily form an instrumental belief that pressing a lever brings about access to food. If the contingency changes so that the rats gain access to food independently of pressing the lever, rats adjust to the new contingency immediately and stop pressing the lever. They have, therefore, revised their belief system to achieve a better fit in the mind-to-world direction of fit. Presumably they are able to do this because they have cognitive mechanisms that can detect when a mismatch between their mental states and their environment occurs. But as I already noted, none of this requires that rats need to employ reflective activities that involve a conceptual understanding of beliefs.

While rats and other mammals may use belief revision strategies that are below the conceptual level, this does not obviously preclude other creatures using conceptual abilities to revise their stock of beliefs. Their conceptual understanding of belief may emerge as a consequence of consciously reflecting upon their self-corrective behaviour in belief revision, thereby making explicit what they only implicitly knew when revising their beliefs. Once the explicit or conceptual understanding of belief is reached, revising one's beliefs may become more sophisticated and flexible. But with the introduction of non-conceptually based abilities that grasp the contrast between true belief and false beliefs, Davidson's argument leading from the having of beliefs to the presupposed mastery of the concept of belief crumbles.

Notice a further interesting implication of the alternative position relevant to the discussion of Davidson's argument. If we allow the possibility that some animals develop their concepts of belief by consciously reflecting upon their self-correcting behaviour, then those animals may build a concept of belief in the absence of other organisms; this means without interacting with other group-members as the interpreter has to do with his subject. Consciously reflecting upon their *own* self-correcting behaviour in an attempt to make sense of them need not necessarily involve interaction with other con-specifics in the group. But once this is granted, Davidson's argument leading him from the concept of belief to the context of interpretation is put into doubt. The alternative position, therefore, allows animals to build a concept of belief in the absence of an interpretation of other organisms of its group.

To summarise: the evaluation of Davidson's arguments has shown that (P1) of the argument is far from being uncontroversial. I have outlined an alternative position that does not conclude that the ability to grasp the distinction between true belief and false belief necessarily involves the concept of belief. A further implication of the alternative position is that even if such a concept of belief is required to grasp the distinction between true and false belief, Davidson's conclusion still would not follow. The concept of belief could also arise in contexts that do not depend on interactions with other group members, especially not in contexts of interpretation. Thus we can conclude that Davidson's argument is far from being either uncontroversial or sound. Davidson cannot conclusively rule out that animals have beliefs.

3. Conclusion

We started the chapter with the presentation of Davidson's argument, which wants to prove that animals do not share mental states with human beings. If true, Davidson's argument could unsettle the belief in a cognitive continuum that connects animals with humans. Put differently, if an abyss were to open up between animals and humans on account of their possession of mental states, then the evolutionary projects of understanding mental states would be seriously weakened.

The basic strategy of the argument against animal belief consists in uncovering special features of mental states that make it virtually impossible for nonhuman animals to attain anything remotely approximating mental states.

Firstly, Dennett holds that he has found such a special feature in the intentional nature of mental states. The intentional nature of thoughts sanctions the fine-grained distinctions between mental states that we are accustomed to draw. For Davidson, however, these distinctions are possible only on the basis of language. I introduced considerations against Davidson's points by showing how we are still able to make sense of attributing determinate mental content to nonhuman animals even in the absence of language. These considerations showed that animals are able to use their own stock of concepts on the basis of which they are able to draw the fine-grained distinctions between their mental states.

Secondly, Dennett argues that the having of beliefs requires the possession of a concept of *belief,* which, in turn, only makes sense in the context of linguistic interpretation. Against this point, I argued that Davidson uses an unsubstantiated premise in his argument. The contrast between true and false beliefs does not demand conceptual abilities for belief revision, as Davidson seems to think. I showed instead that nonhuman animals may use non-conceptual methods to master the distinction between true and false beliefs. In addition, these non-conceptual methods can be used in the absence of interpretative processes that engage in interactions with other group-members.

Davidson's argument fails because he does not engage systematically with alternative accounts of the observations he raises. I presented briefly, for instance, how modular cognitive architectures could answer some of Davidson's observations in nonhuman animals. The possibility of such alternative accounts ensures that we are not conceptually confused when we attribute mental states to animals. Thus animal mental states are not just a convenient fiction we use in order to explain or predict their behaviour, these mental states are psychologically real in the case of animals. With this last conclusion we can demonstrate that evolutionary projects such as evolutionary taxonomies of cognitive capacities are legitimate projects that we should pursue further in our studies of the mind.

Concluding Remarks

We have now seen that evolutionary psychology provides an account of the human mind that differs considerably from the account typically advocated in traditional cognitive psychology. According to evolutionary psychology, the human mind is an ancient engineering product that was invented by evolution to promote the survival or reproduction of our ancestors in their past environments. To this end, evolution equipped the mind with large amounts of specialised cognitive capacities that are dedicated to solving specific adaptive problems our ancestors repeatedly encountered in their physical and social world. Thus the mind is virtually brimming with specialised cognitive equipment that evolution fashioned in order to control our behaviour, and there is also a degree to which natural selection among our ancestors still patterns our current behaviour. This is because evolution entrusted the blueprints of the human mind's cognitive architecture to our genes, which then express the cognitive architecture as required.

This evolutionary picture of the human mind is diametrically opposed to the picture typically favoured in traditional cognitive psychology and other social sciences. There we find the prominent assumption that the human mind is able to reinvent itself anew through culture in a manner unconstrained by its evolutionary heritage. It might be that evolution equips the human infant with a small number of basic and general instincts in the beginning, but it is held more generally that the evolution of a culture-absorbing mind like ours was accompanied by the erasing of specialised cognitive circuitry. Thus the cognitive architectures of human beings that traditional cognitive psychology and other social sciences study are to a large extent the products of culture.

The differences between evolutionary psychology and traditional cognitive psychology have led to healthy debates about the correct ways to approach the study of the human mind. Many objections have been levelled against evolutionary psychology, but the centre of the debate focuses on the status of general-purpose cognitive capacities.

While traditional cognitive psychologists maintain that general-purpose cognitive capacities are the default cognitive option of the mind's cognitive structure, evolutionary psychologists argue instead that general-purpose cognitive capacities can only be the exception to the rule once an evolutionary perspective has been taken into account. General-purpose cognitive capacities are thus inherently suspect, and require evolutionary justification before they can be included as proper parts of the mind's cognitive furniture. I predict that once the dust has settled around the on-going debates about the core concepts of evolutionary psychology and its theoretical support, the evolutionary status of general-purpose cognitive capacities will occupy centre stage. Systematic studies into the status of these cognitive capacities will greatly further our understanding of the evolution of cognition.

In this thesis, I critically introduced and evaluated evolutionary psychology with a view to the evolutionary status of general-purpose cognitive capacities. My evaluation found that there are different versions of evolutionary psychology that are directly relevant to the question about the evolutionary status of general-purpose cognitive capacities. I distinguished between strong versions of evolutionary psychology that exclude general-purpose cognitive capacities, and weak versions of evolutionary psychology that allow the general-purpose cognitive capacities to play a vital role in the cognitive architecture of the mind – provided that these general-purpose cognitive capacities can work with the massively modular cognitive machinery that the mind also possesses and an evolutionary account can be given of them. My thesis demonstrated that weak versions of evolutionary psychology can indeed provide a consistent and fruitful research programme for the study of the human mind.

I accomplished this as follows: Chapter I introduced evolutionary psychology on the basis of its core concepts and theoretical arguments, and chapter II presented a series of objections against evolutionary psychology and defended weak versions of evolutionary psychology against all of them. Chapter III then analysed the question of whether weak versions of evolutionary psychology could develop evolutionary accounts of the origin of general-purpose cognitive capacities. To achieve this, I explored in detail one basic idea: namely, that the massively modular mind confronts specific problems because of the simultaneous activation of different cognitive modules. The chapter thus examined the various conditions under which evolution could have selected for general-purpose cognitive capacities. Chapter IV analysed the core assumption that weak versions of evolutionary psychology bring to the evolutionary study of the human mind: the assumption that non-human animals possess cognitive capacities. The chapter analysed and defended an evolutionary taxonomy of cognitive capacities in non-human animals by critically engaging with the evolutionary taxonomy developed by Dennett. Chapter V assessed arguments from philosophy that seemingly show that projects such as the evolutionary taxonomy proposed in the previous chapter are not as clear-cut as they seem. The chapter defended evolutionary taxonomies by showing that these arguments are not conclusive.

The weak version of evolutionary psychology developed in my thesis is thus able to deal successfully with the most important counter-arguments. Together with the evolutionary account of general-purpose cognitive capacities developed in chapter III, the evolutionary taxonomy of cognitive capacities developed in chapter IV, and the rejection of arguments from philosophy against such evolutionary taxonomies developed in chapter V, weak evolutionary psychology provides a promising research strategy into the mind's structure that should be explored further by anyone interested in the mind. Evolutionary psychology can thus potentially provide interesting and substantive answers to many problems scientists now face in such diverse fields as psychiatry, ethics, social psychology, economics, politics, and cross-cultural studies. Promising first steps are underway in applying the evolutionary perspective in these fields, and it is to be expected that the results will be interesting and surprising. Finally, as with all emerging and young sciences, we should also give evolutionary psychology's research programme a real chance. We should therefore explore the possibilities provided by evolutionary psychology until either evolutionary psychology runs into contradictory empirical findings that ultimately force its downfall, or it proves that it can indeed provide substantial insights to our understanding of the human mind.

References

Alexander, R.D. [1974]: "The evolution of social behaviour", *Annual Review of Ecology and Systematics,* 5, pp. 325–383.

Alexander, R.D. [1979]: *Darwinism and Human Affairs,* Seattle, University of Washington Press.

Alexander, R.D. [1990]: "Epigenetic rules and Darwinian Algorithms", *Ethology and Sociobiology,* 11, pp. 241–303.

Allen, C. [1992]: "Mental Content", *British Journal for the Philosophy of Science,* 43, pp. 537–553.

Allen, C. [1999]: "Animal concepts revisited: the use of self–monitoring as an empirical approach", *Erkenntnis,* 51 (1), pp. 33–40.

Allen, C. and Bekoff, M. [1999]: *Species of Mind,* Cambridge MA, MIT Press.

Atran, S. [1990]: *The Cognitive Foundations of Natural History,* New York, Cambridge University Press.

Ayala, F.J. [1982]: *Population and Evolutionary Genetics: A Primer,* The Benjamin/Cummings Publishing Company, Inc.

Baillargeon, R. [1986]: "Representing the Existence and the Location of Hidden Objects: Object Permanence in 6- and 8-Month-Old Infants", *Cognition,* 23, pp. 21–41.

Balleine, B.W. and Dickinson, A. [1991]: "Instrumental performance following reinforcer devaluation depends upon incentive learning", *Quarterly Journal of Experimental Psychology,* 43B, pp. 279–96.

Balleine, B.W. and Dickinson, A. [1998]: "Consciousness – the Interface between Affect and Cognition", in J. Cornwell *(ed.), Consciousness and Human Identity,* Oxford, Oxford University Press, 1998.

Barkow, J., Cosmides, L., and Tooby, J. *(eds.)* [1992]: *The Adapted Mind: Evolutionary Psychology and the Generation of Culture,* New York, Oxford University Press.
Baron-Cohen, S. [1995]: *Mindblindness: An Essay on Autism and Theory of Mind*, Cambridge MA, MIT Press.
Bellugi, U., Wang, P.P., Jernigan, T.L. [1994]: "Williams Syndrome: An Unusual Neuropsychological Profile", in S.H. Broman and J. Grafman *(eds.)*, *Atypical Cognitive Deficits in Developmental Disorders: Implications for Brain Function,* Hillsdale NJ, Erlbaum, pp. 23–56.
Betzig, L. *(ed)* [1997]: *Human Nature: A Critical Reader*, New York, Oxford University Press.
Block, N. [1986]: "Advertisement for a semantics for psychology", *Midwest Studies in Philosophy,* 10.
Block, N. [1987]: "Functional role and truth conditions", *Proceedings of the Aristotelian Society,* LXI, pp. 157–181.
Boghossian, P.A. [1994]: "Inferential-Role Semantics and the Analytic/Synthetic Distinction", *Philosophical Studies*, 73, pp. 109–122.
Botterill, G. and Carruthers, P. [1999]: *The Philosophy of Psychology,* Cambridge University Press.
Buss, D. M. [1995]: "Evolutionary psychology: A new paradigm for psychological science", *Psychological Inquiry,* 6, pp. 1–30.
Buss, D.M. [1999]: *Evolutionary Psychology: A New Science of the Mind,* Boston, Allyn and Bacon, 1999.
Byrne, R.W. [1996]: *The Thinking Ape Evolutionary Origins of Intelligence*, Oxford University Press.
Byrne, R.W., Corp, N. and Byrne, J.M.E. [2001]: "Estimating the complexity of animal behaviour: How mountain gorillas eat thistles", *Behaviour,* 138, pp. 525–557.
Byrne, R.W. and Whiten, A. *(eds.)* [1988]: *Machiavellian Intelligence,* Oxford University Press.
Byrne, R. W. and Whiten, A. [1992]: "Cognitive Evolution in Primates: Evidence from tactical deception", *Man,* **27**, pp. 609–627.

Carey, S., Spelke, E. [1994]: "Domain-Specific Knowledge and Conceptual Change", in L. Hirschfeld and S. Gelman *(eds.), Mapping the Mind,* Cambridge, Cambridge University Press, pp. 169–200.

Carruthers, P. [1992]: *The Animal Issue: Moral Theory in Practice,* Cambridge University Press.

Chagnon, N.A. and Irons, W. *(eds.)* [1979]: *Evolutionary Biology and Human Social Behavior: An Anthropological Perspective*, North Scituate MA, Duxbury Press.

Chomsky, N. [1988]: *Language and Problems of Knowledge,* MIT Press.

Cosmides, L. [1989]: "The Logic of Social Exchange: Has Natural Selection Shaped How Humans Reason?", *Cognition,* 31, pp. 187–276.

Cosmides, L. and Tooby, J. [1987]: "From Evolution to Behavior: Evolutionary Psychology as the Missing Link", in J. Dupré *(ed.), The Latest on the Best: Essays on Evolution and Optimality*, Cambridge MA, MIT Press, pp. 277–306.

Cosmides, L. and Tooby, J. [1989]: "Evolutionary Psychology and the Generation of Culture, Part II: Case Study: A Computational Theory of Social Exchange", *Ethology and Sociobiology,* 10, pp. 51–98.

Cosmides, L. and Tooby, J. [1992]: "Cognitive Adaptations to Social Exchange", in J. Barkow, L. Cosmides, and J. Tooby *(eds.), The Adapted Mind: Evolutionary Psychology and the Generation of Culture*, New York, Oxford University Press.

Cosmides, L. and Tooby, J. [1994a]: "Beyond Intuition and Instinct Blindness: Toward an Evolutionarily Rigorous Cognitive Science", *Cognition,* 50, pp. 41–77.

Cosmides, L. and Tooby, J. [1994b]: "Origins of Domain-Specificity: The Evolution of Functional Organization", in S. Gelman and L. Hirshfeld *(eds.), Mapping the Mind: Domain Specificity in Cognition and Culture,* New York, Cambridge University Press.

Cosmides, L. and Tooby, J. [1999]: "Unravelling the Enigma of Human Intelligence: Evolutionary Psychology and the

Multimodular Mind", in R.J. Sternberg and J.C. Kaufman *(eds.), The Evolution of Intelligence*, Hillsdale, NJ, Erlbaum.
Cosmides, L. and Tooby, J. [2000]: "Evolutionary Psychology and the Emotions", in M. Lewis and J.M. Haviland-Jones *(eds.), Handbook of Emotions*, New York, Guilford.
Crain, S. and Pietroski, P. [2001]: "Nature, Nurture and Universal Grammar", *Linguistics and Philosophy*, 24, pp. 139–185.
Crane, T. [1990]: "The Language of Thought: No Syntax without Semantics", *Mind and Language,* 5 (3), pp. 187–212.
Crawford, C. [1998]: "Environment and Adaptations: Then and Now", in C. Crawford and D.L. Krebs *(eds.), Handbook of Evolutionary Psychology*, Laurence Erlbaum Assoc.
Cummins, R. [1975]: "Functional Analysis", *Journal of Philosophy*, 72, pp. 741–65.
Cummins, R. [1983]: *The Nature of Psychological Explanation*, Cambridge MA, Bradford Books/The MIT Press.
Davidson, D. [1980]: *Essays on Actions and Events*, Oxford, Oxford University Press.
Davidson, D. [1982]: "Rational Animals", *Dialectica,* 36, pp. 317–28.
Davidson, D. [1984]: "Thought and Talk", in D. Davidson, *Inquires into Truth and Interpretation,* New York, Oxford University Press.
Davidson, D. [1984a]: *Inquiring into Truth and Interpretation,* New York, Oxford University Press.
Davidson, D. [1997]: "Seeing Through Language", in J. Preston *(ed.), Thought and Language,* Cambridge, Cambridge University Press.
Davidson, D. [1999]: "The Emergence of Thought", *Erkenntnis,* 51, pp. 7–17.
Davis, J. and Bitterman, M.E. [1971]: "Differential reinforcement of other behavior (DRO): A yoked-control comparison", *Journal of the Experimental Analysis of Behavior,* 15, pp. 237–41.
Davies, P.S. [1999]: "The Conflict of Evolutionary Psychology", in V. Hardcastle *(ed.), Where Biology Meets Psychology*, MIT Press, pp. 67–81.

Dawkins, R. [1986]: *The Blind Watchmaker*, New York, Norton.
Dawkins, R. [1989]: *The Selfish Gene*, New York, Oxford University Press.
Dennett, D.C. [1978]: *Brainstorms: Philosophical Essays on Mind and Psychology*, Bradford Books.
Dennett, D.C. [1987]: *The Intentional Stance*, Cambridge, MIT Press/ A Bradford Book.
Dennett, D.C. [1994]: "The Role of Language in Intelligence", in J. Khalfa *(ed.)*, *What is Intelligence?*, Cambridge, Cambridge University Press.
Dennett, D.C. [1995]: *Darwin's Dangerous Idea*, New York, Simon & Schuster.
Dennett, D.C. [1996]: *Kinds of Minds: Towards An Understanding of Consciousness*, New York, Harper Collins Publishers.
Devitt, M. [1996]: *Coming to Our Senses*, New York, Cambridge University Press.
Dickinson, A. [1980]: *Contemporary Animal Learning Theory*, Cambridge, Cambridge University Press.
Dickinson, A. [1988]: "Intentionality in Animal Conditioning", in L. Weiskrantz *(ed.)*, *Thought without Language*, Oxford, Clarendon Press.
Dickinson, A. [2001]: "Causal Learning: An Associative Analysis", *The Quarterly Journal of Experimental Psychology*, 54B(1), pp. 3–25.
Dickinson, A. and Balleine, B.W. [1993]: "Actions and Responses: The Dual Psychology of Behaviour", in N. Eilan, R. McCarthy, and M.W. Brewer *(eds.)*, *Spatial Representation*, Oxford, Blackwell, 1993.
Dickinson, A. Balleine, B.W. [1994]: "Motivational Control of Goal-Directed Action", *Animal Learning and Behavior*, 22, pp. 1–18.
Dickinson, A. and Balleine, B.W. [1999]: "Causal Cognition and Goal-Directed Action", in C. Heyes and L. Huber *(eds.)*, *The Evolution of Cognition*, MIT Press.

Dunbar, R.I.M. [1993]: "Coevolution of neocortical size, group size and language in humans", *Behavioral and Brain Sciences,* 16, pp. 681–735.
Dunlap, K. [1912]: "The case against introspection", *Psychological Review*, 19, pp. 404–413.
Dupré, J. [1994]: "The Mental Lives of Nonhuman Animals", in M. Bekoff and D. Jamieson *(eds.), Readings in Animal Cognition,* Cambridge MA, MIT Press.
Elman, J., Bates, E., Johnson, M, Karmiloff-Smith, A., Parisi, D., and Plunkett, K. [1996]: *Rethinking Innateness: A Connectionist Perspective on Development*, Cambridge, MA, MIT Press.
Field, H. [1977]: "Logic, meaning, and conceptual role", *Journal of Philosophy,* 69, pp. 379–408.
Field, H. [1978]: "Mental Representation", *Erkenntnis*, 13, pp. 9–61.
Fodor, J. [1975]: *The Language of Thought*, New York, Crowell.
Fodor, J. [1983]: *The Modularity of Mind: An Essay on Faculty Psychology*, Cambridge MA, MIT Press.
Fodor, J. [1987]: *Psychosemantics: The Problem of Meaning in the Philosophy of Mind*, Cambridge Ma, MIT Press.
Fodor, J. [2000]: *The Mind Doesn't Work That Way: The Scope and Limits of Computational Psychology*, Cambridge MA, MIT Press.
Fodor, J. and LePore, E. [1992]: *Holism: A Shoper's Guide,* Oxford, Blackwell.
Fodor, J. and McLaughlin, B. [1990]: "Connectionism and the Problem of Systematicity: Why Smolensky's Solution Doesn't Work", *Cognition,* 35, 183–204.
Fodor, J. and Pylyshyn, Z. [1988]: "Connectionism and the Cognitive Architecture: A Critical Analysis", *Cognition,* 28, pp. 3–71.
Frith, U. [1989]: *Autism*, Blackwell.
Gallistel, C.R. [1990]: *The Organization of Learning*, Cambridge MA, Bradford Books/MIT Press.
Gallistel, C.R. *(ed.)* [1992]: *Animal Cognition, Cognition Special Issues,* Cambridge MA, MIT Press/Bradford Book.

Garfield, J.L. *(ed.)* [1989]: *Modularity in Knowledge Representation and Natural-Language Understanding*, Cambridge MA, MIT Press.

Gazzaniga, M.S. and Altman, J.S. *(eds.)* [1998]: *Brain and Mind: Evolutionary Perspectives*, The Human Frontier Science Program.

Gelman, R. and Baillargeon, R. [1983]: "A Review of Some Piagetian Concepts", in J.H. Flavell and E.M. Markman *(eds.)*, *Handbook of Child Psychology*, New York, Wiley, pp. 167–230.

Gigerenzer, G. [1997]: "The modularity of social intelligence", in A. Whiten and R.W. Byrne *(eds.)*, *Machiavellian Intelligence II: Extensions and Evaluations*, Cambridge University Press.

Godfrey-Smith, P. [1999]: "Environmental Complexity and the Evolution of Cognition", in R.J. Sternberg and J.C. Kaufman *(eds.)*, *The Evolution of Intelligence*, Hillsdale NJ, Lawrence Erlbaum.

Gopnik, A. and Meltzoff, A. [1997]: *Words, Thoughts, and Theories*, Cambridge MA, MIT Press.

Gopnik, A. and Wellman, H. [1994]: "The Theory Theory", in L. Hirshfeld and S. Gelman *(eds.)*, *Mapping the Mind: Domain Specificity in Cognition and Culture*, Cambridge University Press, pp. 257–93.

Gould, J. L. and Gould, C. G. [1994]: *The Animal Mind*, New York, Scientific American Library.

Gould, S.J. [1997a]: "Darwinian Fundamentalism", *The New York Review of Books*, June 12.

Gould, S.J. [1997b]: "Evolution: The Pleasure of Pluralism", *The New York Review of Books*, June 26, pp. 34–37.

Gould, S.J. [2000]: "More Things in Heaven and Earth", in H. Rose and S. Rose *(eds.)*, *Alas, Poor Darwin*, London: Jonathan Cape, pp. 85–106.

Gould, S.J. and Lewontin, R. [1979]: "The Spandrels of St Marco and the Panglossian Paradigm", *Proceedings of the Royal Society of London B*, 205, pp. 581–598.

Gould, S.J. and Vrba, E.S. [1982]: "Exaptation – a Missing Term in the Science of Form", *Paleobiology*, 8(1), pp. 4–15.

Grantham, T. and Nichols, S. [1999]: "Evolutionary Psychology: Ultimate Explanations and Panglossian Prediction", in V. Hardcastle *(ed.), Where Biology Meets Psychology*, MIT Press, pp. 47–66.

Gregory, R.L. [1981]: *Mind in Science: A History of Explanations in Psychology*, Cambridge, Cambridge University Press.

Griffin, D.R. [1976]: *The Question of Animal Awareness: Evolutionary Continuity of Mental Experience*, New York, Rockefeller University Press.

Griffin, D.R. [1984]: *Animal Thinking*, Harvard University Press.

Griffin, D.R. [1992]: *Animal Minds*, University of Chicago Press.

Hamilton, W. D. [1964]: "The genetical evolution of social behavior", *Journal of Theoretical Biology*, 7, pp. 1–52.

Harman, G. [1973]: *Thought*, Princeton, Princeton University Press.

Harman, G. [1987]: "(non-solipsistic) Conceptual Role Semantics", in E. LePore *(ed.), New Directions in Semantics*, London, Academic Press.

Hauser, M. [1996]: *The Evolution of Communication*, MIT Press.

Hauser, M. [1998]: "Report: A nonhuman primate's expectations about object motion and destination: The importance of self-propelled movement and animacy", *Developmental Science*, 1(1), pp. 31–37.

Heyes, C. and Dickinson, A. [1990]: "The Intentionality of Animal Action", *Mind and Language*, 5(1).

Heyes, C. and Huber, L. [2000]: *The Evolution of Cognition*, The MIT Press.

Hirschfeld, L. and Gelman, R. *(eds.)* [1994]: *Mapping the Mind: Domain-Specificity in Cognition and Culture*, Cambridge, Cambridge University Press.

Holland, P. [1979]: "Differential Effects of Omission Contingencies on Various Components of Pavlovian Appetitive Conditioning", *Journal of Experimental Psychology: Animal Behavior Processes*, 5.

Holland, P. [1992]: "Event Representation in Pavlovian Conditioning: Image and Action", in C.R. Gallistel *(ed.), Animal Cognition*, MIT Press.
Holland, P. and Straub, J.J. [1979]: "Differential Effects of Two Ways of Devaluing the Unconditional Stimulus after Pavlovian Appetitive Conditioning", *Journal of Experimental Psychology: Animal Behavior Processes*, 5.
Hull, C.L. [1943]: *Principles of Behavior: An Introduction to Behavior Theory*, New York, Appleton Century Crofts.
Hume, D. [1973]: *Inquiry Concerning Human Understanding*, New York, Bobbs Merrill.
Humphrey, N. K. [1976]: "The social function of intellect", in P. P. G. Bateson and R. H. Hinde *(eds.), Growing points in ethnology*, Cambridge, Cambridge University Press.
Irons, W. [1979]: "Natural Selection, Adaptation, and Human Social Behavior", in N.A. Chagnon and W. Irons *(eds.), Evolutionary Biology and Human Social Behavior*, Duxbury Press.
Irons, W. [1990]: "Let's Make Our Perspective Broader Rather Than Narrower", *Ethology and Sociobiology*, 11, pp. 361–74.
Johnson, M. and Morton, J. [1991]: *Biology and Cognitive Development: The Case of Face Recognition*, Blackwell.
Karmiloff-Smith, A. [1992]: *Beyond Modularity: A Developmental Perspective on Cognitive Science*, MIT Press.
Karmiloff-Smith, A. [2000]: "Why Babies' Brains are not Swiss Army Knives", in H. Rose and S. Rose *(eds.), Alas, Poor Darwin*, London, Jonathan Cape, pp. 85–106.
Krebs, J.R. and N.B. Davies, N.B. [1993]: *An Introduction to Behavioural Ecology*, Blackwell Science.
Laurence, S. and Margolis, E. [2001]: "The Poverty of the Stimulus Argument", *British Journal for the Philosophy of Science*, 52(2), pp. 217–276.
Lehrman, D.S. [1964]: "The reproductive behavior of ring doves", *Scientific American, 211 (6), pp. 48–54.*
Leslie, A.M. [1987]: "Pretence and Representation: The Origins of 'Theory of Mind'", *Psychological Review*, 94, pp. 412–26.

Leslie, A.M. [1994]: "ToMM, ToBy, and Agency: Core Architecture and Domain-Specificity", in L. Hirschfeld and S. Gelman *(eds.), Mapping the Mind: Domain-Specificity in Cognition and Culture,* Cambridge, Cambridge University Press.
Leslie, A.M. [1995]: "Pretending and believing: Issues in the theory of ToMM", *Cognition*, 50, pp. 193–220.
Luca Cavalli-Sforza, L. [1997]: "Genes, peoples, and languages", *Proceedings of the National Academy of Sciences*, 94, pp. 7719–7724.
Malcolm, N. [1977]: "Thoughtless Brutes", in N. Malcolm, *Thought and Knowledge: Essays by Norman Malcolm*, Ithaca NY, Cornell University Press.
Marler, P. [1970]: "A comparative approach to vocal learning: Song development in white-crowned sparrows", *Journal of Comparative Psychology and Physiology*, 71, pp. 1–25.
Marr, D. [1982]: *Vision: A Computational Investigation into the Human Representation and Processing of Visual Information*, San Francisco, Freeman.
McFarland, D. [1989]: *Problems of Animal Behaviour*, Essex, Longman Scientific&Technical.
Miller, G.F. [2000a]: *The Mating Mind: How Sexual Choice Shaped the Evolution of Human Nature*, London: William Heinemann.
Miller, G.F. [2000b]: "Mental Traits as Fitness Indicators: Expanding Evolutionary Psychology's Adaptationism", in D. LeCroy and P. Moller *(eds.), Evolutionary Approaches to Human Reproductive Behavior*, Anls. NY Acad. Sci. 907.
Millikan, R.G. [1984]: *Language, Thoughts, and Other Biological Categories*, MIT Press.
Millikan, R.G. [1993]: *White Queen Psychology and Other Essays for Alice*, MIT Press.
Mithen, S. [1996]: *The Prehistory of the Mind: A Search for the Origins of Art, Religion and Science*, London, Phoenix.
Murphy, D. and Stich, S. [2000]: "Darwin in the madhouse: Evolutionary psychology and the classification of mental disorders", in P. Carruthers and A. Chamberlain *(eds.), Evolution and the Human Mind,* Cambridge University Press, pp. 62–92.

O'Leary-Hawthorne, J. [1993]: "Belief and Behavior", *Mind and Language*, 8, pp. 461–86.
Park, M.A. [1996]: *Biological Anthropology*, Mayfield Publishing Company.
Pavlov, I. [1927]: *Conditioned Reflexes*, Oxford, Oxford University Press.
Peacocke, C. [1992]: *A Theory of Concepts*, Cambridge MA, MIT Press.
Pearce, J.M. [1987]: *An Introduction to Animal Cognition*, Laurence Erlbaum Publisher.
Pearce, J.M. [1997]: *Animal Learning and Cognition*, Psychology Press.
Piaget, J. [1971]: *Biology and Knowledge*, Chicago, University of Chicago Press.
Pinker, S. [1996]: *How the Mind Works*, New York, Norton.
Pinker, S. and Bloom, A. [1990]: "Natural Language and Natural Selection", *Behavioral and Brain Sciences*, no. 13, pp. 707–84.
Profet, M. [1992]: "Pregnancy Sickness as Adaptation: A Deterrent to Maternal Ingestion of Teratogens", in J. Barkow, L. Cosmides, and J. Tooby *(eds.), The Adapted Mind: Evolutionary Psychology and the Generation of Culture*, New York, Oxford University Press, pp. 327–365.
Pyke, G.H. [1984]: "Optimal Foraging Theory: A Critical Review", *Annual Review of Ecology and Systematics*, Vol. 15, pp. 523–575.
Quine, W.V. [1960]: *Word and Object*, Cambridge MA, MIT Press.
Ramachandran, V.S. and Blakeslee, S. [1998]: *Phantoms in the Brain: Probing the Mysteries of the Human Mind*, New York, Quill William Morrow.
Rashotte, M.E., Griffin, R.W., and Sisk, C.L. [1977]: "Second-Order Conditioning of the Pigeon's Key Peck", *Animal Learning and Behavior*, 5, pp. 25–38.
Rescorla, R.A. and Wagner, A.R. [1972]: "A Theory of Pavlovian Conditioning: Variation in the Effectiveness of Reinforcement and Nonreinforcement", in A.H. Black and W.F. Prokasy *(eds.), Classical Conditioning II: Current Research and Theory*, New York, Appleton-Century-Crofts, pp. 64–99.

Ristau, C. *(ed.)* [1991]: *Cognitive Ethology: The Minds of Other Animals*, Hillsdale, NJ: Lawrence Erlbaum.
Ristau, C. [1991]: "Aspects of the Cognitive Ethology of an Injury-Feigning Bird, the Piping Plover", in C. Ristau *(ed.), Cognitive Ethology*, Hillsdale, NJ, Laurence Erlbaum.
Rizley, R.C. and Rescorla, R.A. [1972]: "Associations in Second-Order Conditioning and Sensory Preconditioning", *Journal of Comparative and Physiological Psychology*, 81.
Rose, H. and Rose, R. [2000] *Alas, Poor Darwin*, London: Jonathan Cape.
Russon, A.E., Bard, K.A., and Taylor Parker, S *(eds.)* [1996]: *Reaching Into Thought: The Minds of the Great Apes*, New York, Cambridge University Press.
Samuels, R. [1998]: "Evolutionary Psychology and the Massive Modularity Hypothesis", *British Journal for the Philosophy of Science*, 49, pp. 575–602.
Samuels, R. [2000]: "Massive Modular Minds: Evolutionary Psychology and Cognitive Architecture", in P. Carruthers and A. Chamberlain *(eds.), Evolution and the Human Mind: Modularity, Language and Meta-Cognition*, Cambridge, Cambridge University Press.
Schiffer, S. [1987]: *Remnants of Meaning*, Cambridge MA, MIT Press.
Scholl, B.J. and Leslie, A.M. [1999]: "Modularity, Development and 'Theory of Mind'", *Mind and Language,* 14(1), pp. 131–53.
Searle, J. [1994]: "Animal Minds", *Midwest Studies in Philosophy*, XIX, Blackwell Publishing.
Segal, G. [1996]: "The Modularity of Theory of Mind", in P. Carruthers and P. Smith *(eds.), Theories of Theories of Mind*, Cambridge, Cambridge University Press.
Shepard, R. N. [1984]: "Ecological constraints in internal representations", *Psychological Review*, 91(4), pp. 417–447.
Shepard, R. N. [2001]: "Perceptual-Cognitive Universals as Reflections of the World", *Behavioral and Brain Sciences,* 24.
Sherry, D. and Schacter, D. [1987]: "The Evolution of Multiple Memory Systems", *Psychological Review*, 94, pp. 439–54.

Shettleworth, S.J. [1998]: *Cognition, evolution, and behavior*, New York, Oxford University Press.

Shettleworth, S.J. [2001]: "Animal Cognition and Animal Behaviour", *Animal Behaviour,* 61, pp. 277–86.

Skinner, B.F. [1938]: *The Behavior of Organisms: An Experimental Analysis*, New York, Appleton Century Crofts.

Slater, A. and Kirby, R. [1998]: "Innate and learned perceptual abilities in the newborn infant", *Experimental Brain Research,* Vol. 123, pp. 90–94.

Slater, A. and Quinn, P.C. [2001]: "Face recognition in the newborn infant", *Infant and Child Development, special issue on Face Perception and Recognition,* 10, pp. 21–24.

Smith, N.V. and Tsimpli, I.M. [1995]: *The Mind of a Savant: Language Learning and Modularity*, Oxford, Blackwell.

Sober, E. [1997]: "Morgan's Canon", in D.D. Cummins and C. Allen *(eds.), The Evolution of Mind,* Oxford, Oxford University Press.

Spelke, E. [1988]: "The Origins of Physical Knowledge", in L. Weiskrantz *(ed.), Thought without Language*, Oxford, Clarendon Press, 1988.

Spelke, E. [1990]: "Principles of Object Perception", *Cognitive Science*, 14, pp. 29–56.

Spelke, E. [1994]: "Initial Knowledge: Six Suggestions", *Cognition*, 50, pp. 431–45.

Sperber, D. [1994]: "The modularity of thought and the epidemicological of representations", in L. Hirshfeld and S. Gelman *(eds.), Mapping the Mind: Domain-Specificity in Cognition and Culture*, New York, Cambridge University Press.

Sperber, D. and Wilson, D. [1986]: *Relevance: Communication and Cognition*, Blackwell.

Sterelny, K. [1990]: *The Representational Theory of Mind: An Introduction*, Blackwell Publisher.

Stich, S. [1978]: "Beliefs and subdoxastic states", *Philosophy of Science*, 45(4), pp. 499–518.

Stich, S. [1983]: *From Folk Psychology to Cognitive Science*. Cambridge MA, MIT Press.
Stotz, K.C., and Griffiths, P.E. [2001]: "Dancing in the Dark: Evolutionary Psychology and the Argument from Design", in S. Scher and M. Rauscher *(eds.), Evolutionary Psychology: Alternative Approaches*, Dordrecht: Kluwer.
Symons, D. [1990]: "Adaptiveness and Adaptation", *Ethology and Sociobiology*, 11, pp. 427–44.
Symons, D. [1992]: "On the Use and Misuse of Darwinism in the Study of Human Behavior", in J. Barkow, L. Cosmides, and J. Tooby *(eds.), The Adapted Mind: Evolutionary Psychology and the Generation of Culture*, New York, Oxford University Press.
Thorndike, E.L. [1911]: *Animal Intelligence: Experimental Studies*, New York, Macmillan.
Tinbergen, N. [1951]: *The Study of Instinct*, Oxford, Oxford University Press.
Titchener, E.B. [1912]: "The Schema of Introspection", *American Journal of Psychology*, 23, pp. 485–508.
Tolman, E.C. [1932]: *Purposive Behavior in Animals and Men*, New York, Century.
Tolman, E.C. [1948]: "Cognitive Maps in Rats and Men", *The Psychological Review*, 55(4), pp. 189–208.
Tomasello, M. and Call, J. [1997]: *Primate Cognition*, Oxford, Oxford University Press.
Tooby, J. and Cosmides, L. [1990]: "The Past Explains the Present: Emotional Adaptations and the Structure of Ancestral Environments", *Ethology and Sociobiology*, 11, pp. 375–424.
Tooby, J. and Cosmides, L. [1992]: "The Psychological Foundation of Culture", in J. Barkow, L. Cosmides, J. Tooby *(eds.), The Adapted Mind: Evolutionary Psychology and the Generation of Culture*, New York, Oxford University Press.
Tooby, J. and Cosmides, L. [1997]: "Letter to the Editor of the New York Review of Books", *The New York Review of Books*, July 7.

Tooby, J. and Cosmides, L. [1998]: "Evolutionizing the Cognitive Sciences: A Reply to Shapiro and Epstein", *Mind and Language*, 2, pp. 195–204.
Turke, P.W. [1990a]: "Which Humans Behave Adaptively, and Does It Matter?", *Ethology and Sociobiology*, 11, pp. 305–39.
Turke, P.W. [1990b]: "Which Humans Behave Adaptively, and Does It Matter?", *Ethology and Sociobiology*, 11, pp. 445–463.
Vauclair, J. [1996]: *Animal Cognition: An Introduction to Modern Comparative Psychology*, Harvard University Press.
Vygotsky, L.S. and Luria, A.R. [1993]: *Studies on the History of Behaviour: Ape, Primitive, and Child*, V.I. Golod and J.E. Knox *(eds.)*, Hillsdale NJ, Erlbaum.
Wagner, G.P. [1995a]: "Adaptation and the Modular Design of Organisms", in F. Moran, A. Moreno, J.J. Merelo, and P. Chacon *(eds.), Lecture Notes in Artificial Intelligence: Advances in Artificial Life*, Heidelberg, Springer-Verlag, pp. 317–28.
Wagner, G.P. [1995b]: "Homologues, Natural Kinds, and the Evolution of Modularity", *American Zoologist*, 36, pp. 36–43.
Wagner, G.P. and Altenberg, L. [1996]: "Complex Adaptations and the Evolution of Evolvability", *Evolution*, 50, pp. 967–76.
Wagner, G. P., Mezey, J. & Calabretta, R. [2001]: "Natural Selection and the origin of modules", in W. Callabaut, and D. Rasskin-Gutman *(eds.), Modularity. Understanding the development and evolution of complex natural systems*, MIT Press.
Watson, J.B. [1913]: "Psychology as the Behaviorist views it", *Psychological Review*, 20, pp. 158–177.
Wellman, H. [1990]: *The Child's Theory of Mind*, Cambridge MA, MIT Press.
Whiten, A. and Byrne, R.W. *(eds.), Machiavellian Intelligence II: Extensions and Evaluations,* Cambridge University Press.
Williams, G.C. [1966]: *Adaptation and Natural Selection*, Princeton University Press.
Wilson, E.O. [1975]: *Sociobiology: The New Synthesis*, Cambridge MA, Harvard University Press.
Wundt, W. [1897]: *Outlines of Psychology*, Wilhelm Engelmann.

Zahavi, A., Zahavi, A. and Balaban, A. [1997]: *The Handicap Principle*, Oxford University Press.

FSC
www.fsc.org
MIX
Papier | Fördert
gute Waldnutzung
FSC® C083411

Kolibri 360 GmbH
Ferdinand-Jühlke-Straße 7
99095 Erfurt, Deutschland
produktsicherheit@kolibri360.de

Druck:
CPI Druckdienstleistungen GmbH
im Auftrag der
Kolibri 360 GmbH
Ferdinand-Jühlke-Str. 7
99095 Erfurt